FOREST URBANISMS

NEW NON-HUMAN AND HUMAN ECOLOGIES
FOR THE 21ST CENTURY

LEUVEN UNIVERSITY
PRESS

EDITED BY
BRUNO DE MEULDER,
KELLY SHANNON

LAP

The peer-reviewed series LAP (*Landscape and Architecture Projections*) focuses on design research in the fields of architecture, urbanism and landscape. It seeks to highlight innovative practices worldwide which boldly address the most pressing socio-political, ecological and spatial issues of the contemporary times.
It emphasizes work which is developed in cooperation with activists and civil society, various governmental and/or development agencies and stakeholders as well as with other experts.

Series Editors
- Kelly Shannon
- Ward Verbakel

Editorial Board
- Eliana Barbosa, The Federal University of Rio de Janeiro
- Margarita Jover Biboum, Tulane University / aldayjover architecture and landscape
- Ilze Wolff, Wolff Architects

CONTENTS

FOREWORD — 7
PREFACE — 11

FRAMING FOREST URBANISMS — 15
A TWENTY-FIRST CENTURY AGENDA
Bruno De Meulder, Kelly Shannon

FORESTS & SCIENCE — 75
I. THE 3 + 30 + 300 RULE FOR URBAN FORESTRY — 76
Chiara Cavalieri, Cecil Konijnendijk
II. ON THE NEED OF LARGE OLD TREES TO KEEP CITIES YOUNG AND VIBRANT — 82
Rik De Vreese, Bart Muys
III. INTENTIONALLY INCLUSIVE URBAN FORESTRY — 88
Colleen Murphy-Dunning

FOREST URBANISMS PROJECTS — 97
I. LANDSCAPE AFTER NATURE — 98
Bureau Bas Smets
II. FISH TAIL PARK — 104
Kongjian Yu, Turenscape
III. DEN HOUT 2040 — 110
Wim Wambecq, Joris Moonen, MIDI
IV. SALVADOR ETHNOBOTANICAL GARDEN — 116
Embyá Paisagens & Ecossistemas
V. 'NATURE VILLAGE' — 120
EFFEKT

VI. TRANSFORMING SOUTHBANK BOULEVARD 126
  TCL

VII. MADRID METROPOLITAN FOREST (ZONE 4) 132
  aldayjover architecture and landscape

FOREST URBANISMS EXPLORATIONS 141

I. NATIVE, EXOTIC OR FUTURE-PROOF? NAVIGATING URBAN TREE PLANTING IN THE BRUSSELS MAELBEEK VALLEY 142
  Björn Bracke, Koenraad Danneels, Marlène Boura

II. FORESTING A *CHAR* IN THE BRAHMAPUTRA VALLEY IN ASSAM, INDIA 157
  Swagata Das, Kelly Shannon, Bruno De Meulder

III. VISIBLE AND INVISIBLE FORESTS. THE CULTIVATION OF SHADE IN WINNIPEG, MANITOBA, CANADA 171
  Kamni Gill

IV. BRIDGING GREEN GAPS. EMPOWERING PARTICIPATORY GOVERNANCE THROUGH TREE PLANTING IN BARRANQUILLA, COLOMBIA 184
  Alejandra Parra-Ortiz, Gina Serrano-Aragundi

V. URBAN FORESTS AS POST-MANICURE OUTDOOR DESIGN TYPES 200
  Jörg Rekittke

VI. A FENCE THAT GREW A FOREST. A STRATEGY FOR A PARK AT PACHACAMAC ARCHAEOLOGICAL SANCTUARY 213
  Takako Tajima

VII. FOREST LOGICS, LENSES AND ORDERS. TOWARDS A CLIMATE-FORWARD FOREST URBANISM ALONG THE ERIE CANAL NATIONAL HERITAGE CORRIDOR 228
  Jamie Vanucchi, Maria Goula

ILLUSTRATION CREDITS 249

COLOPHON 252

FOREWORD
FOREST URBANISMS AS A POLITICAL PROJECT

Cities and trees share a long and contrasted history. For a very long time, at least in the European context, trees appeared as singularities in a mostly mineral urban space. Streets were not planted, and parks were nonexistent. When they were present, trees were to be found in private settings: in the kitchen gardens, orchards, and pleasure gardens of aristocratic residences and convents. However, things began to change during the second half of the seventeenth century, with the development of tree-lined avenues and boulevards at the periphery of the urban cores and the opening to the public of royal gardens such as Hyde Park in London or the Tuileries in Paris. The eighteenth century found it self-evident that nature was beneficial in physical as well as in moral terms. Trees in particular appeared in an eminently positive light. The stroll under their canopy was considered as healthy by doctors and social reformers alike. From the nineteenth century onwards, planners began to consider trees as a global system, thus paving the way for the contemporary assimilation of all urban trees as a forest, a forest that is both discontinuous at certain points and continuous at others. With the rise of the regional dimension, such a system appeared more and more as a structuring feature of the rapidly growing cities of the industrial age (Picon 2024).

Despite this rich legacy – still present in the double set of virtues (some physical, others moral) that we attribute to trees – something has definitely changed in our approach to their role in urban settings. More generally, the role of nature in cities is evolving rapidly under the pressure of climate change and the need to adapt cities to extreme weather events, from heatwaves to droughts or floods. Focusing on trees draws conclusions from this shift as fundamental, perhaps as the cumulated effect of all that happened in prior centuries.

To begin with, the systemic dimension of the presence of trees in cities that appeared at the end of the nineteenth century has become more vital and above all more complex. The aim is no longer to simply plant and ensure the development of selected species in a relatively hostile urban environment characterized by a significant level of pollution. Rather, it is about promoting a biodiversity that extends far beyond what traditional tree linings, gardens, and parks allowed. This goal entails finding new and creative equilibria between a planned nature based on carefully chosen specimens and an unplanned, often unruly nature that often comprises invasive species. Certainly, the reflections of environmentalists and the work of landscape architects such as Gilles Clément, the promoter of the notion of the "planetary garden", have paved the way for the realization that we need to expand our vision beyond the tame arrangements of avenues and parks. The challenges, however, are no less daunting.

How can we expand our horizon beyond the traditional figures of the tree line, the garden, and the park in the traditional sense? In a global age, marked by a growing awareness of the diversity of cultures, new possibilities arise from the study of non-Western traditions ranging from indigenous clearings to Asian sacred groves. The present book offers stimulating reflections on what can be gained from these alternative models and practices. It also reviews different urban situations that reinforce this need for an innovative approach to nature in cities, to urban forests especially. Forest urbanisms: the plural to the term urbanism reinforces the call for a diversity of approaches, though they share a common ground with the concern to give a new importance to trees.

The case studies reinforce the urgency to rethink urban planning using nature as a guideline. In such a context, one may be tempted to search for positive rules such as the 3+30+300 one, three trees in sight from home or the workplace, 30% of canopy in every part of the city, and 300 meters from a green space, in an effort to quantify contemporary biophilia and its health and mental components. Useful as they may be, these rules of thumb are only part of the answer, however. As the editors and some of the contributors to this volume rightly point out, practices such as urban agriculture force us to consider more generally alternatives to our dominant modes of urbanization. Rethinking urban planning may entail, among other things, challenging the sharp distinction between the city and the countryside.

What if trees and more general green spaces were as, if not more essential than built objects in the city of the future? In her famous 1958 essay, *The Human Condition*, Hannah Arendt evoked the way humans make a world for themselves through designed objects that stabilize both their environment and their relationships with one another (Arendt 1958). For her, this world built from objects that she characterized as works, as opposed to the transitory productions of labour, made political life possible. Other natural beings played little role in this approach to what is human. That sort of forgetting has become impossible, for we realize day after day that our world would not be possible without our constant interaction with natural elements and beings, from soil to vegetation. What forest urbanisms ultimately propose is to recognize that we also make a world for ourselves, a more and more urban world given the overall pace of urbanization, through our connection with nature.

From such a perspective, the question becomes how to envisage political life differently? Whereas politics has been all about the deliberations and actions of humans, the time may have come to find ways to associate non-humans to the political sphere. But how to give them a voice in the debate? To answer this challenge, Bruno Latour had imagined "parliament of things", which would comprise representatives of rivers, mountains, forests, and of course animal species (Latour 2004). Despite the attempts made here and there to implement such deliberative bodies, this proposal still remains distant.

In the meantime, the forest urbanisms analyzed in this book represent a first step in this direction. Instead of telling us how to act in concert with non-humans, they tell us how to contribute to the construction of worlds shared by human as well as natural elements and beings. Before we can start writing the new political play called for by the Anthropocene, perhaps we should begin by building the theater that will house it: a different kind of city, based on a rediscovered partnership with nature.

July 2024

Antoine Picon
G. Ware Travelstead Professor of the History of Architecture and Technology
Director of Doctoral Programs
Harvard Graduate School of Design

References

Arendt, Hannah. (1958). *The Human Condition.* Chicago, London: The University of Chicago Press.

Latour, Bruno. (2004). *Politics of Nature: How to Bring the Sciences into Democracy.* Cambridge, MA; London: Harvard University Press.

Picon, Antoine. (2024). *Natures Urbaines: Une Histoire Technique et Sociale 1600-2030,* Paris, Pavillon de l'Arsenal.

PREFACE

Forest Urbanisms is one of several offshoots of an interdisciplinary, three-day international conference, *Urban Forestry, Forest Urbanisms & Global Warming,* held in Leuven in June 2022, co-chaired by Cecil Konijnendijk, Chiara Cavalieri, and Kelly Shannon. The event brought together forestry scientists, policy makers, and designers (architects, landscape architects, and urbanists). Many of them cross their respective disciplinary borders: forestry scientists involved in community work, policy makers who interact with designers, and designers collaborating with a host of experts operating in other fields. The exchange revealed the potential as well as difficulty of their distinctive understanding of trees in the built environment and, vice versa, of settlements within forested environments. As Jörg Rekittke commented in a conference review in the *Journal of Landscape Architecture* (JoLA), "there was a palpable tension between scientists, policy experts and designers throughout the conference. However, rather than detracting from the event's success, this diversity of perspectives ultimately contributed to its high academic quality" (Rekittke 2023: 109). All participants acknowledged the *urgency* and *necessity* of renewed nature < > culture relationships, including novel and designed co-existences of human and non-human species. The contemporary cascade of crises – from socio-economic to political to spatial and ecological – demands the radical reconceptualization of the occupation of the earth.

Humanity's role in changing the face of the earth, already scrutinized in the 1950s by numerous scholars (Thomas 1956), needs a draconic revision.

The book, the second in the LAP series, points to an opening towards the serious rethinking of forestry and human settlement. The former continues to disappear at an alarming rate while the latter is increasingly omnipresent, including in the most remote corners of the world. During the neolithic era, a practice was put in motion: settling required clearing (components) of forest or, in arid parts of the world, creating (walled) gardens with ornamental or fruit trees. Whatever the method, settling, originally and for a very long time, implied a fundamental relationship with trees. However, somewhere along the road to 'progress' with industrialization and the unfolding of capitalism, the relationship was dramatically compromised and in many cases completely lost. The co-presence of nature and culture of sorts evaporated. One can wonder, with the omnipresence of settlements, if a reset is needed, that is, complementing the urban everywhere, where appropriate, with afforestation. As far-fetched as it sounds, it would have a guaranteed impact on global warming, from micro to macro scales – assuming the careful stewardship of forests and trees and attending significantly more to ecology. It would set the scene for a different way for humans to occupy the world. It would construct new worlds that encourage more interplay between human and nonhuman species. The conviction of the editors of this volume is that constructing these worlds requires deliberate design of forests and tree stands, so as to guide the development of settlement and change the face of the earth for the better. Hence the conscious inclusion of a section with design projects that demonstrate the capacity of design and which underscore the diversity of scales and approaches found in on-going design research.

July 2024

Bruno De Meulder,
Kelly Shannon

References

Rekittke, Jörg. (2024). Urban Forests, Forest Urbanisms & Global Warming. *Journal of Landscape Architecture*, 18 (1), 109-110.

Thomas, William L. (1956). *Man's Role in Changing the Face of the Earth* [2 vols.]. Chicago, London: National Science Foundation and Wenner-Gren Foundation for Anthropological Research.

FRAMING FOREST URBANISMS

A Twenty-First Century Agenda

BRUNO DE MEULDER,
KELLY SHANNON

FOREST URBANISM: A PROVISIONAL OUTLINE

Humankind's ever-increasing role in changing the face of the earth (Thomas 1956) has led to the enormous socio-ecological crisis of the world. Humanity needs to radically alter the unsustainable way it occupies and transforms the world, as well as how it settles *with* and *within* the world. Urbanism has an evident responsibility in this regard, alongside agriculture and forestry, the two other major domains which radically alter the earth and its ecologies. Since the Industrial Revolution, the development and systematic application of land-use planning in all corners of the globe have continuously defined and redefined coverage and occupation for these three discrete entities. Each has its own governance structure, steered by short-sighted neoliberal policies that overemphasize clarity and productivity and which are informed by separate disciplines (urbanism, agronomy and forestry) operating, so to speak, on their own planet. Disciplinary separation has incited progressive specialization, upscaling both the intensification and commodification of occupation categories. Unsurprisingly, the contemporary multi-faceted socio-ecological crisis strongly resonates in urban, agriculture and forests crises. It is no wonder that attempts are being made to undo the separation. Urban forestry, urban agriculture and agro-forestry are relatively new terms that blur boundaries and attempt to reset dichotomous systems. Urban forestry advocates for the systemic integration of trees and tree stands in urban environments. Similarly, urban agriculture and agroforestry are promoted, albeit with differences. For promoters of vernacular agro-forestry, such as Geneviève Michon, the proposition is not to add a layer of forestry to agriculture (as is too often mechanically applied in Western contexts), but to understand and appreciate vernacular practices of agriculture which are embedded within forests – with the notion forest interpreted in the widest possible way (Michon 2015).

Interestingly, the 'forest' shares with the 'city' the fact that they are both difficult to precisely define. They both may constitute worlds unto themselves, spaces of multiplicity. In

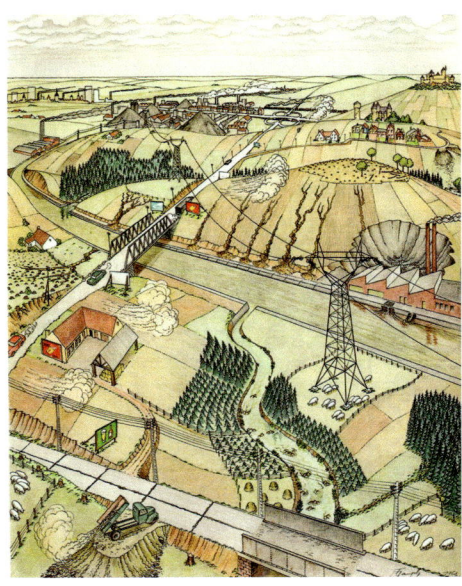

The before (left) and future (right) drawings of Martin Tanghe in the 'Ecological Synthesis' (1980) of Paul Duvigneaud. Duvigneaud was a pioneer of the Brussels school of urban ecology; his work sought a rebalance of urban and natural systems.

conventional forestry science, a rather bureaucratic and arbitrary definition is often used: a minimum surface, coverage and average height (respectively 0.5ha, 10% and 5m in the FAO-definition). This reductive and quantitative approach contrasts strongly with the amazingly extensive and sensitive vocabulary in different languages, contexts and cultures. They depict a wide variety of types and characters of forests, ranging from dense to thin forests, forest gardens and hedge landscapes, boscages, woodlands, constellations of *bosquests*, hunting forest, jungles and so on. Michon replaces quantitative criteria with the qualitative and imbues a wide variety of environments with an enduring and structural role taken up by trees – from the sparsely tree-covered *dehesa* (in Spain) and *montado* (in Portugal) to the world's densest forests (Michon 2015). As differentiated as forms and characters of forests can be, so too are the meanings that trees have – from variations of sacredness (meanings that are eroding quickly everywhere as industrialization and modernization engulf the world (Lemaire 2023)), yet resisting in more resilient cultures), to different forms of productivity (a meaning to which it is reduced in market-driven contexts with predatory extractivist practices), to aesthetic values and, finally, to simply different ecological values (that are increasingly being understood to their full extent) (Simard 2021).

Agro-forestry and urban forestry both reject the categorical segregation of nature (forest) and culture (agriculture and settlement). Similarly, forest urbanism could be understood as an urbanism (the science of the city as well as the discipline that spatially organizes the environment) in which the forest is central: trees and tree communities have structural and enduring roles in the environment. In forest urbanism, the three forms of occupation of the world (forest, agriculture, and settlement) are co-present and seek a new symbiosis. Agriculture and settlement can be embedded within the forest. A link to the sensitive exploration of the environment through the notorious anti-military and anti-colonial science-fiction novel of Ursula K. Le Guin *The Word for World is Forest* (Le Guin 1972) can be made as a strong reminder of how humankind's morality is tied to forests. Le Guin decried brutal environmental violence, the despoiling of natural resources and dehumanizing beyond any measure as nothing more than the private gain for an absolute minority. The recently published *The Word for World is Still Forest* recognized Le Guin's canonical work, while traversing various contemporary woodlands to reveal how practices of care, concern and attention also enable humans to inhabit and flourish in this world as forest (Springer & Turpin 2017).

In forest urbanism, the deeply rooted, albeit artificial, distinction between nature (forest) and culture (settlement and agriculture) of Western thought must be transcended (Descola 2015). If the word for world is forest, it implies that forest urbanism is a world unto itself, operating across scales – from the world to the region, the city, neighbourhood or hamlet, and to the lonely monumental eastern African baobab that stands sentry, rising above savannas, or to the singular building and its interplay with trees. While discussing "tree gardens", Gina Crandell reminds how trees, the "largest living architectural structures", co-mingle with architecture (Crandell 2013: 11). Clearly, forest urbanism is not only about quantities. According to the Greek myth, the goddess Athena planted a single sacred olive tree on the Acropolis (Lemaire

2023:228). Forest urbanism addresses the ecologically battered world with matters of fact (that were often quantified) and matters of concern, as Latour has suggested (Latour 2004), as well as with matters of care (Puig de la Bellacasa 2011). It analyzes, reflects and subsequently intervenes through actions, projects, programmes, plans and policies. Yet also simply by planting a tree, we are planting hope (Lucy Larcom, as cited in Jones 2016: ix), a concrete action in the here-and-now which is for the benefit of future generations and subsequent ecological succession. Forest urbanism does not shy away from immediate, small, concrete actions, but also not from formulating and propagating a *Leitbild*, "guiding images" (Chombart de Lauwe 1964) – the utopian dimension of paradigms of urbanism.

Since the forest and the urban (both with a multiplicity of meaning) cannot be defined unambiguously, forest urbanism does not fit any precise definition, either. What follows in this framing essay are twenty points of orientation, each with a synthesizing text and a related image. Both indicate elements, developments, examples, inspirations and notions of the world of forest urbanism, as it has already existed, implicitly or explicitly, since time immemorial, and as it can be (re)articulated for the twenty-first century. The points incorporate insights and practices from related disciplines and revisit several past paradigms and practices of urbanism. They seek to reverse the deleterious and extractive processes and practices that have been accelerating at alarming rate since the sixteenth century (Grewe & Hölzl 2018). Scientific and technological progress and colonialism created a toxic cocktail with regards to forests and the larger environment. In short, this essay is a provisional opening, a conceptual atlas of forest urbanism. It articulates elements to work with, developments to reverse, knowledge to incorporate, practices to (re-)consider and directions to follow. Hence, the alternation of small texts and images that all are invitations to take note of insights, see things, to then relate them and start imagining over and over again how forest urbanism can contribute to a new world in the making. Or, in the words of Georges Didi-Huberman, to see, collect and narrate (Didi-Huberman 2023). The essay begins with a consideration of forests in the threefold manner in which humans occupy the world (forest, agriculture, urbanism). It concludes with two expectations: (19) a new paradigm of forest urbanism and (20) new morphologies and typologies of forest urbanism. The essay is not a continuous text, but comprised of fragments that weave together notions from various disciplines, across time and geographic locations. This provisional outline of forest urbanisms is followed by three sections: a set of contributions from the sciences, a package of contemporary forest urbanism projects, and a series of essays that detail some aspects of forest urbanism.

1. FORESTS: A SELF-RENEWING PRIMEVAL MATRIX

Forests are a primeval matrix that both preceded and have always been a primary environment and resource for humanity and for the entire non-human world of fauna and flora. They are the oldest complex (heterarchical and dynamic) land-based ecosystems on the planet. As palaeoarchaeologist Patrick Roberts has stated, "in essence they developed an entirely new planetary order … for the emergence and diversification of life on Earth … and key to the evolution of all plant life and to its increasingly permanent seat at the table of atmospheric, geological and climatic change" (Roberts 2021: 19,22). At the same time, particularly in Western thought, they were often regarded as the wild/rough/feral, magical/supernatural/enchanted and dangerous/innocuous/forbidden. Etymologically "forest" is derived from both the Latin *forum* as in "land subject to a ban" and *foris* as in "an outside condition" and "beyond the realm of the enclosed" (Di Palma 2014: 178). It is also related to *silva* and "savage". The word jungle, commonly used for forests, comes from the Hindi *jangal*, classifying a *terra incognita*, something outside the realm of human settlement and home comforts, a place of lawlessness and violence. In the cultural imagination of the West, forests were outside, counter to the civilized and brought into culture (Mulder 2020). Paradoxically, forests have been understood by

A complex tropical forest profile with interwoven forest floors hosting rich biodiversity, as documented by the renowned botanist and biologist Francis Hallé together with a group of students from Montpellier University in French Guiana.

many as unproductive, with uniformly poor soils, deadly hazards, and wild animals. However, in many non-Western contexts, they were regarded as rich environments – though not to say 'treasuries' – with the highest species biodiversity and hosting a multitude of resources (for hunting, collecting/gathering, logging, picking, etc.);

they are a renewable source for building material, nutrition, medicine, fossil-based energy. In many indigenous communities, they remain environments to settle within. For them, the word for world is (still) forest (Springer & Turpin 2017).

Forests have always been more than treasuries. Forest Urbanism inherently relies on spaces of multiplicity: sacred spaces, wild/natural spaces, resource spaces and recreational spaces, amongst others.

2. FORESTS, AGRICULTURE, URBANISM: OCCUPYING THE WORLD IN A RECOVERED TRIALOGUE

The domestication of nature, bringing it into culture – or as the French radical socialist and colonial governor Albert Sarraut would phrase it *"mettre en valeur"* (Sarraut 1923) (bringing into value, developing) – often implied the clearing of forests for agriculture that in its slipstream generated settlement (or vice versa, the settling that requires agriculture to sustain itself). In the West, nature (forests) and culture (agriculture and urbanism) came to be seen equally as opposites, with the latter continuously consuming more and more of the former (Descola 2015). Self-regeneration of forests was often excluded, given that for a very long time, wood was massively used as an essential and primary material (for everyday tools, construction, shipbuilding, weapons, cooking, heating, etc.). 'Cut and move' was far too often the general (careless) practice (Handel 2011). Until the nineteenth century, when steel replaced wood in shipbuilding, most maritime-based powers (republics/kingdoms/nations and empires) had a devastating impact on forests. Hence, the appetite

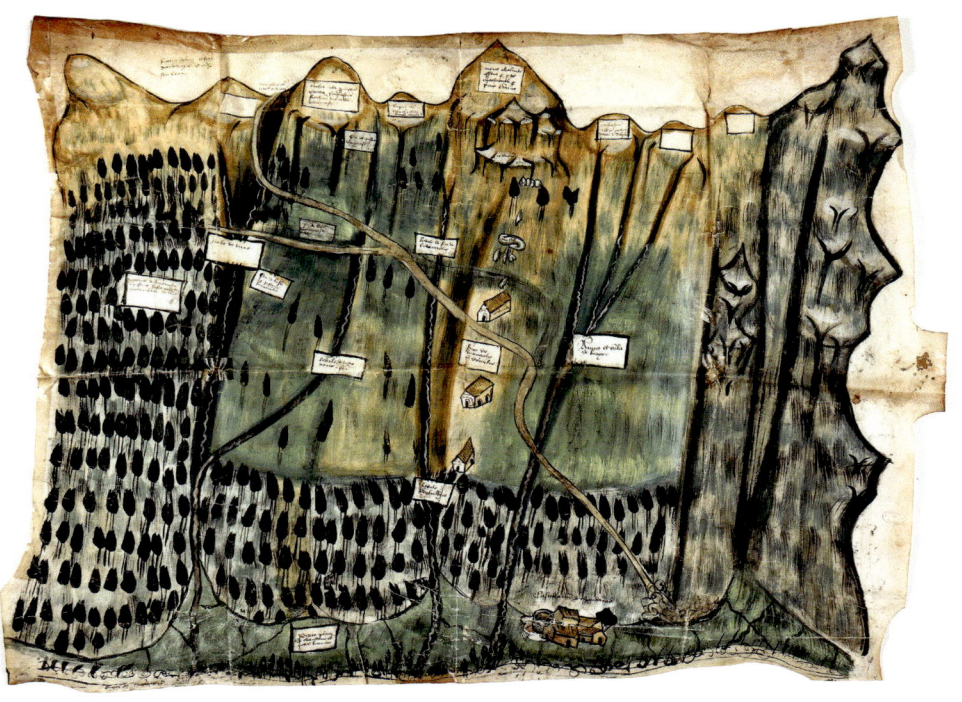

Jean de Briancon's painting of the valley of Casteldelfino in Italy (1422) exemplifies the more than 12,000 years-old trialogue of interdependent functions of the territory (agriculture, forestry and settlement). Courtesy of Archives départementales de l'Isère, B 4496.

for colonies that were, since antiquity, "established in forested regions to assure the founding cities' timber supplies" (Hughes & Thirwood 1982: 70). Besides military and commercial shipbuilding, cities and mining were *de facto* the largest forest destroyers. Observers such as George Perkin Marsh asserted that city-states of antiquity and subsequent eras "consumed their own future as they felled their forests and allowed their soil to wash away" (as quoted in Hughes, Thirwood 1982: 60). The consequences of the systemic destruction of 'woods' and a subsequent plea for conservation strategies was already a central issue in George Perkin Marsh's *Man and Nature* (1864). Paraphrasing the Old Testament while making sharp factual observations, Marsh stated that

> Man has too long forgotten that the earth was given to him for usufruct alone, not for consumption, still less for profligate waste. Nature has provided against the absolute destruction of any of her elementary matter…

But she has left it within the power of man irreparably to derange the combinations of inorganic matter and of organic life (Marsh 1864: 34).

In *Taking the Country's Side: Agriculture and Architecture*, Sébastian Marot insightfully and refreshingly appeals to ending the separation between agriculture and settlement that were once intrinsically linked out of necessity, yet lost ties in the industrial era of urban growth, market development, technological evolution and specialization (Marot 2019). It is a necessary to forcefully and explicitly bring forestry back into this equation, moving from the disrupted and human-centric dialogue between urbanism and agriculture to a dynamically balanced trialogue between urbanism, agriculture and forests. Evidently, forests were initially co-present, as agriculture and settlements were carved out of the forests by clearing them (often considered as a catalyst for many civilizations) (Michon 2015: 45). It is not just co-presence, however, that matters. Regardless of the colonial appetite of Greek city-states, Aristotle did not argue by accident that an "ideal city would have its own forest near enough to ensure self-sufficiency" (Hughes & Thirwood 1982: 62). Unfortunately, until now, short-term gains prevail for greedy (neo-)colonial endeavours. Whether understood through the culture/nature dialectic or otherwise, a symbiotic trialogue between forest, agriculture and settlement is as natural as can be. In some rural cultures, settlement and agriculture remain embedded within (different degrees of domesticated) forests. For ages, however – and unfortunately – increasingly and in too many contexts, forestry, agriculture and settlement have been radically segregated, as stated above. The industrial era, with its monofunctional planning obsession, took such categorization to an extreme. The current ecological crisis requires an increase and omnipresence of forests (that contain the largest biodiversity). Forests within and around the city, forestry combined with agriculture, yet simultaneously also as an element in itself: forests as forest *tout court* should be on the agenda everywhere, as Francis Hallé has advocated (Hallé 2021). Forests necessarily and structurally contribute to the half-earth thesis of E.O. Wilson, where Hallé's proposal to re-create a primeval forest in the heart of western Europe is part and parcel of such an endeavour, one in which local communities are to be integrated (Wilson 2016).

Forest Urbanism is a plea for a constructive trialogue of forestry, agriculture and urbanism. It implies the abandonment of the categorical zoning of the world. It advocates a necessary multiplicity that forestry, agriculture and urbanism all (need to) have.

3. FOREST CLEARINGS AND WALLED GARDENS: ARCHETYPES OF LANDSCAPE ARCHITECTURE AND SETTLEMENT

Christophe Girot claims that "landscape is evidence of an accrued intelligence of place through topological transformation and an exchange of techniques, beliefs, and actions" (Girot 2016: 15–16). The marking of ground (Gregotti 1996), of differentiating cared-for societal space from the *as-found* untended natural environment, was a primordial activity of humankind. For Girot, forest clearing and the walled garden are two enduring archetypes of (Western) landscape architecture. Forest clearings – both sacred (ritual and mortuary purposes) and profane (settlement systems) – evolved from dense forests. They are part and parcel of humanity's quest for shelter, sustenance and safety. Later, clearings also became constructs, with trees planted around a void. The paradigmatic 'primitive hut' of abbé Marc-Antoine Laugier – based on humankind's need to shelter himself from nature – was illustrated by Charles Eisen as a structure comprised of trees and branches within a small clearing in the forest (Laugier 1755). Clearings were equated with

Photographer Vincent Rosenblatt captured the relatively isolated forest clearing settlements of Brazil's Yanamoni.

commons, and the land beyond was *terra nullius* (a no-man's-land) (Di Palma 2014). Conversely, walled gardens – stylized versions of nature – were exclusive and introvert. They were constructed worlds of "selection, collection and propagation" of particular species where the "nature beyond" the garden remained without selection or manipulation (Girot 2016: 30). Trees regulated their microclimates and defined their usability in arid and semi-arid regions (like the Middle East or the Indian Subcontinent). Forest clearings (in the 'wild/natural' environment) and walled gardens (stylizing nature in inhospitable landscapes) created microclimatically moderated and symbolic environments in which to dwell. They also dramatically altered ecologies. The extents of forests were dramatically diminished, and cultural landscapes took precedence. Long ago, clearings were openings within the forest; today, in far too many contexts, forests are the exceptional masses in the openness of the countryside. Yet forests, forest remnants and trees continue to embody deep sentiment for populations throughout the globe, and forest clearings and walled gardens remain archetypes. For Canadian academic Erin Manning, the clearing is not a mere physical space, but about the "violence of cultural clearing and the genocide it leaves behind" (Manning 2023: 10).

Forest Urbanism can make critical use of the archetypes of the forest clearing and walled garden, while recognizing them as settler-colonial acts of defining a territory and producing borders.

4. INDIGENOUS FOREST COMMUNITIES: STEWARDS OF A COMPLEX ECOLOGY

Since time immemorial, there have been communities that live in forests. Indigenous peoples have inalienable rights to their ancestral territories and live both *from* and *with* forest ecologies, all regulated by diverse and context-specific traditional and customary systems of use. Traditional forest knowledge systems and practices are inseparable from the notion of territoriality (International Alliance of Indigenous-Tribal Peoples of the Tropical Forests 1996). Over time, forest-dependent communities have holistically co-evolved with forests and retain a vital relation between humans and non-human species. On the one hand, their shelter, sustenance and livelihoods are derived from the forest. On the other hand, they steward the forest, protecting and enhancing its biodiversity. Forests were modified to increase the fertility of the soil and favour particular types of vegetation through sophisticated vernacular forms of agro-forestry (Michon 2015). Forests are fundamental to Indigenous peoples' sociocultural, spiritual, economic and political survival. Beyond functional meanings, they embody fundamental

Abel Rodriguez's *Chagra Cycle* (2013) depicts how the co-evolution of forests and the forest-dependent indigenous communities sustains a vital relation between humans and non-human species. Forests are their shelters, and they are forests' stewards. Courtesy of Tropenbos Colombia.

symbolic values (Michon 2015). Indigenous forest communities are front-line defenders of invaluable territories. While Indigenous peoples have lived with and to a certain extent domesticated forests since time immemorial, there has been a Western critique of modern life centred on life in the woods as a resistive alternative, as evinced by Henry David Thoreau's foundational text *Walden* (1854), establishing a norm for civil disobedience related to dwelling environments (Thoreau 1954). The forest as a refuge and place of freedom crosses geographies and cultures and is nearly universal – evinced from the Africans enslaved in Brazil who escaped and founded *quilombos* (hinterland settlements) in (adjacent) forests (Tavares 2017) to groups in the mountainous jungles of Southeast Asia that resisted being integrated in state systems (of the lowlands) and the exploitations that go with that (taxation, compulsory crops)

(Scott 2009). More and more, forests are inhabited by new groups, determined to escape the devastating consequences of market economy, adding to the wave of contestations of forest territories, by indigenous people and by social and ecological movements (Vidalou 2017; Lambert 2023). In the 1950s, Pierre de Schlippe (and later, in the 1960s, Colin Turnbull) noted how the shifting cultivation practices of the Zande in Congo shaped the tropical forest through intergenerational cycles of moving, settling temporarily, clearing, hunting, gathering, planting, moving, leaving, or abandoning temporarily, regeneration and returning eventually to "recovered sites" that had largely become palm fruit forests (de Schlippe 1956; Turnbull 1961). Michon describes how members of the Karen ethnic group in Thailand (who define themselves as "children of the forest") systematically domesticate the forest, whereby the forest is understood as an environment that obviously comprises its vegetation and animals, not to mention, of course, humans. "The world and the life of the Karen are governed by a principle of harmony: harmony between men and the forest, harmony between men and spirits" (Michon 2015: 65), which implies the preservation of all the resources of the territory and a fundamental respect for the forest, whereby a reciprocal relationship is developed.

Forest Urbanism can develop novel forms of governance and management which learn from the way indigenous forest communities have been environmental protectors since time immemorial, stewarding landscape connectivity and biodiversity.

5. SACRED GROVES AND FORESTS

For numerous indigenous communities, nature is culture. Forests are not only endowed with tangible resources, but also intangible (spiritual, sociocultural and mental health) properties. Indigenous knowledge and practices are open invitations to rethink humankind's relationship with the living and the material world (Albert & Kopenawa 2022). For Philippe Descola, such an invitation is "a project of repopulating the social sciences with nonhuman beings, and thus of shifting the focus away from the internal analysis of social conventions and institutions and towards the interactions of humans with (and between) animals, plants, physical processes, artifacts, images, and other forms of beings" (Descola 2015). For Eduardo Kohn, the habitat of this "ecology of minds" is, not unsurprisingly, the forests (Kohn 2013). Ancient settlement practices were often choreographed in relation to an intelligent accommodation to the environment and a layered understanding of nature and beliefs in land gods, river kings and forest spirits. In many worldviews, sacred groves and forests were, and still are, places of retreat and pilgrimage, burial and ancestral worship, communal gatherings and festivals, places for the cultivation of medicinal plants, healing, and ritual cleansing. They also are often revered for their watershed value, namely erosion control and maintenance of water quality. In many cultures and communities, individual trees, sacred groves and forest patches are fiercely protected and celebrated through rituals, poetry, literature and sacred texts, legends and myths, and folk songs (Trosper & Parrotta 2012; Pearce 2023). Farms in many contexts were not accidentally accompanied by mature trees (Lemaire 2023). In contexts worldwide, forest spirits are both practical and mystical and have a predominant role in both traditional religious and spiritual beliefs as well as for sociocultural identity. In Asia, cities and dispersed rural settlements historically developed in relation to a worldview that included geomancy (*feng shui*) and divination, which choreographed the activities of humankind within nature. Throughout cities in Japan, trees and "eternal forests" are venerated in relation to *kodama* (folkloric tree spirits*)* and Shintoism (Moore

In the midst of an almost completely deforested Ethiopia are churches with their sacred pocket forests. As captured by Kieran Dodds, the sacred domains contrast sharply with the surrounding secular agricultural landscape.

& Atherton 2020). Similarly, in Cambodia, both trees and water bodies are worshipped in relation to *neak ta* and Buddhism (Edwards 2008). Traditional Khmer settlements are structured by "wats", forested domains that contain pagodas and other religious buildings – all in a precisely defined relationship to topography (De Meulder & Shannon 2023). Traditional settlements, as exemplified throughout the Himalayas, are often structured between temples, water and sacred forests (Anthwal et al. 2010). Elsewhere, from Ethiopia's church forests to Native American sites to pre-Christian pagan and animist places throughout Europe to various Amazonian groups, indigenous peoples are considered traditional custodians of their sacred forests and groves, practicing a form of community-based stewardship of natural resources (Wild & McLeod 2008). They are often associated with a legacy concerning both taboos and customs.

Forest Urbanism can encompass old and new sacred groves and forests. Historic ones are gaining the attention of scientists and conservationists as catalysts for land protection and even restoration. Relic forests are often the only remaining areas of climax vegetation, and their boundaries stand in stark contrast with the secular world.

6. FORESTS: THE EDGE OF CIVILIZATION

In line with many authors, Robert Pogue Harrison, professor of literature, asserts that Western civilization "literally cleared its way in the midst of forests. A sylvan fringe of darkness defined the limits of its cultivation, the margins of its cities, the boundaries of its institutional domain; but also the extravagance of its imagination" (Harrison 1993: ix). The forest changed in its status as a savage wasteland into a cultured domain. Yet for Harrison, forests retain a central place in the human imagination and defy usual civilizational binaries. Expanding on Harrison, the Welsh geographer Michael Williams commented that "forests upset, confuse and destabilize civilization" (Williams 2008). In the process of civilizing, however, the first and last victim has always been the forest. "As civilization becomes centre, the edge recedes so that forests move further away from habitation and the unknown … the forest marks the edge of both the literal and imaginative domains, which we have lost; the outsideness is gone, the edge of exteriority vanishes, and the inside becomes emptier" (Williams 2008: 359). For Ursula K. Le Guin, humankind's morality is tied to forests. She wrote a speculative fiction novella, *The Word for World is Forest,* in 1972 as a direct response to the geopolitical climate and environmental violence of

Continuous and systemic deforestation of the Amazon, as depicted by Clemente Juliuz in 2018, sadly illustrates an enduring attitude towards the forest in too many parts of the world — as a brutal and expendable edge of civilization. Courtesy of Artes Vivas Collection, Verena Regehr.

the American war in Vietnam (Le Guin 1972). In a new edition of the book from 1980, she said,

> The lies and hypocrisies redoubled; so did the killing. Moreover, it was becoming clear that the ethic which approved the defoliation of forests and grainlands and the murder of non-combatants in the name of "peace" was only a corollary of the ethic which permits the despoliation of natural resources for private profit or the GNP, and the murder of the creatures of the Earth in the name of "man." The victory of the ethic of exploitation, in all societies, seemed as inevitable as it was disastrous (quoted in Springer & Turpin 2017: xiv).

American academic Vittoria Di Palma has written extensively on the dual notion of 'wasteland'. On the one hand, it is a desolate or unoccupied land which is yet unmodified by civilization. On the other hand, it is land that is spoiled, consumed, over-actively domesticated or treated with negligence (Di Palma 2014: 3, 186). Throughout the globe, by the eighteenth century, there were great enclosure operations, where forests, heathlands, swamps and other wilderness areas were privatized and "brought into culture" (agriculture) or turned into productive forests. An "ideology of improvement" was almost universally applied to land deemed waste in both its original, as-found state but also that laid to waste by human activity. In the typology of the forest was "the emergence of a division between terms of wilderness and wasteland, with wilderness understood as the fragile, untouched, uninhabited landscape, and wasteland the despoiled landscape, ravaged by the detrimental activities of an unscrupulous culture" (Di Palma 2014: 229).

Forest Urbanism can become imbued with novel relationships that eliminate binaries and create environments for humans and non-human species to thrive and co-exist.

7. URBAN FOREST (PRIVATE) DOMAINS

When forests were first settled, they were community property, governed by customary laws. They were embedded into everyday sociocultural and productive practices; they provided settlers with land to create clearings for habitation as well as fields for pasturage and agriculture (Hardin 1968). They were also the scene of vernacular forms of agro-forestry which were embedded within vast forests. The forests provided fodder for cattle, wood for fuel, and material for construction, as well as medical and edible plants

The notorious *Hunt in the Forest* by Uccello (+/- 1465) depicts the ritual of hunting as a favorite aristocratic pastime of the ancient regime. The enormous forest canopy symbolizes the subjugation of the wild by the nobility (self-proclaimed representatives of civilization).

(Escobar 2008; Michon 2015). However, from the early Middle Ages, in Europe, forest domains were often under the reign of kings, dukes and various royal custody (Di Palma 2014). A nascent form of forest urbanism originated from the numerous castle and abbey domains that were inserted within the once endless forest; besides the castles and abbeys as such, they developed churches, homesteads and outbuildings; initiated hamlets; and founded villages. They were also the domains for hunting, and large tracts of forests were reserved for such exclusive use by nobles. Kings were responsible for protecting their subjects, which included the extermination of forest beasts; the duty was eventually transformed into one of pleasure hunting (Di Palma 2014: 182), and royal privilege was often pitted against traditional common rights (Di Palma 2014: 184). Fast forward: some cities are today lucky enough to be in the vicinity of these domains, which became public (such as the Sonian Forest next to Brussels) after the French Revolution and are nowadays the backbone of the ecological structure of their respective metropoles (De Meulder et al. 2019). Other royal capitals, such as Karlsruhe or Versailles, were planned as a configuration of forest and city (with their palaces at the fulcrum between the two). Regardless, over time, forests became more and more demarcated and regulated; there were clear contestations with regards to use and 'ownership'. The so-called enclosure movement (abolishing traditional user-rights and defining private ownership), which raged through western Europe in the eighteenth and nineteenth centuries, was a point of inflection in this respect. During the nineteenth and twentieth centuries, European colonies went through similar processes that have often been perpetuated by postcolonial nation-states. A chain of successive ownership, management practices, uses, abuses and transformations unfolded, whereby large parts of forests turned from sites of resource extraction (hunting, wood logging, wood products harvesting, etc.) to colonization. On a different note, medieval abbeys largely paved the way for the agricultural exploitation of cleared forest pockets in the domains that they were granted by nobility or royalty. Abbeys were demarcated by walled gardens, the other archetype of landscape architecture besides the forest clearing around a settlement (Girot 2016: 125). In the notorious, though unrealized St. Gall Abbey, the paradigmatic convent plan can also read as a prototype of a nucleated settlement with streets, residential and other buildings, gardens and orchards (Pregill & Volkman 1993: 153–155); in short, it embodied two spatial organizations for an ideal community (Abel 2017). Abbeys quickly experienced the need to manage the triple functions for their granted domains: forests, agriculture and settlements. Each occupation had its own challenges, yet also had to be adapted in relation to one another in order to sustain the self-sufficiency of such largely autarchic communities. When it came to clearing forests, for instance, the monks could not simply rely on the previously mentioned 'cut and move' practice, since their domains were ultimately limited. Hence, out of necessity, forest management was developed, and reforestation practices were experimented with. In this way, forestry expertise was generated in medieval abbeys. Confronted with massive wood consumption, deforestation and soil erosion, royal decrees and governments initiated reforestation campaigns and eventually supported the development of scientific forestry. For centuries, forests have been planned and systematically exploited and maintained; often this management was more extensive and sophisticated than town planning during the same era (Bridel 1798).

Forest Urbanism can re-establish urban forest domains as new commons, replete with everyday practices and new customary laws, while re-establishing the integration of forestry, agriculture and settlement planning together with their respective management practices.

8. SCIENTIFIC FORESTRY

Forestry evolved with the increased management of natural resources. Already in the sixteenth century, Germany initiated a systematic and modern management of forests, in relation to clear-cutting for agriculture and for harvesting timber (Grewe & Hölzl 2018). The three principles of Germany forestry were minimum diversity (to avoid "arbitrary" details of nature and add to forestry efficiency); a balance sheet (dependent on mathematical utilitarianism and "reason" for accurate inventory and accounting, as well as keeping equilibrium in a "forest use budget"); and sustained yield (a dogmatic and bureaucratic annual cycle to "deliver the greatest possible constant volume of wood") (Lowood 1990: 336–8). German forestry produced monocultural, even-aged forests that conformed to the "neatly arranged constructs of science" (Lowood 1990: 340–1). Formal forestry schools were established in both Germany and France in 1825. During the Renaissance, forestry had already been developed as resource management by the Venetian Republic, while the early nineteenth-century plundering of forests by the British in India similarly led to the understanding that 'cut and move' practices resulted in a devastating dead end. By the mid-nineteenth century, scientific forestry became irrevocably tied to colonialism and the careless pillaging and, subsequently,

US Forest Service scientists in 1905 at work near Canton, Mississippi, systematically noting, documenting and interpreting the elements and dimensions of the forest, all through the lens of its utilitarian management.

the consistent and proven necessary management of environments for resources. It was part and parcel of various forms of spatial strategies of occupation and colonization, political violence, socio-cultural obliteration, slavery, debt-peonage and land-grabbing. Sir Dietrich Brandis, a German who introduced forest management in Burma and India, developed the notion of prescriptive rights to forest users, in addition to systematic and centralized (state) management through long-term planning – all for the sake of revenue yields. Trees were regarded as mere material supply and logging concessions led to a progressive detachment from the land – through neutral and efficient expertise. Technological advances not only optimized extraction but also regeneration processes.

Over time, colonial governments, followed by the modern state and large extractive companies, took possession over more and more forests and imposed centralized and technocratic management, with economic efficiency driven regimes of monocultural planting, fire suppression, selective thinning and harvesting by clear-cutting (Worster 1993). Belief in supposedly rational technique trumped all else, and forests became a crop to be nurtured for sustained yields. Natural forest dynamics were completely altered and there was habitat destruction and massive biodiversity loss; forests underwent a process of ecological simplification. An elite and engineered form of environmental imperialism came together, as colonial knowledge prevailed through the control of forests. At the same time, there has been resistance by local communities and the forest itself. Paradoxically, scientific forestry was also carried out in the name of land and forest conservation. Gifford Pinchot (US Forest Commission and US Forest Service, 1896-1910) was infamous for his development of policies which sought to retain, conserve and manage public land through the means of scientific forestry (Miller 2004). According to Carrere and Lohmann, there is a new "forestry imperialism" related to the vast paper industry that exploits vast industrial tree plantations of monocultures (of conifers, eucalyptus, acacia, and others) in the Southern hemisphere; hopefully, the digital revolution will be a saviour to such a devastating process (Carrere & Lohmann 1996).

Forest Urbanism can utilize the ever-evolving science of forestry but keep it in check with a value system which favours renaturalization and a productivity that does not devastate ecologies, while respecting rights of local communities.

9. GARDEN CITIES: INHABITED FORESTS

In the twentieth century, cities developed an explicit urban forest relationship, particularly in early 'garden city' experiments. In Ebenezer Howard's 1898 foundational diagram for the Garden City, new forests appear in the four corners of the new settlements' hinterland, as cornerstones for the whole plan. The diagram balances the three functions of agriculture, settlement and forests, while emphasizing the need for planted forests during the process of urbanization. Howard's garden city was "a peaceful path to real reform", a paradigmatic shift in planning that tackled the dreadful social and ecological conditions of the nineteenth-century industrial city (Howard 1898). The creation of these new forests was also an essential part of Howard's proposition. Since then, variations of the Garden City movement have left a built legacy of canonical forest urbanism projects across the globe, including Forest Hills Gardens in Queens, New York (1909, by Frederick Law Olmsted Jr. and Grosvenor Atterbury); Jardim América in São Paulo in Brazil (1913, by Barry Parker and Raymond Unwin); Canberra Forest City in Australia (1914, by Walter Barry Griffith); Sokol Plan in Moscow (1923 by Nikolay Markovnikov, Alexi Shchusev and others); Waldseidlung Onkel Toms Hütte in Berlin (1926–1932, by Bruno Taut, Hugo Haring and Otto Rudolf Salvisberg); and Sunila in Finland (1936–1938, 1947, 1951–1954 by Alvar Aalto). Tapiola, the first post-war new town in Finland remains, despite recent waves of densification and development, an iconic manifestation of forest urbanism, where forest elements give shape to the city across the spectrum (Von Hertzen & Spreiregen 1974). New housing typologies were conceived in direct relation with the forested environment in which they were embedded. A social reform agenda (which often gained momentum in times of crisis, such as post-war reconstruction) drove many of the projects, while others were developed as leafy suburbs for elites (Ward 1992). Particularly in Scandinavia, experiments that interweave forest typologies and urban fabrics remain a living tradition (as exemplified by the project of Effekt, in the Forest Urbanism Projects section of this

In archetypal medieval settlement structures, agriculture, forestry and settlement were interwoven, as exemplified in this early fifteenth-century miniature from the chapter "Les pays de la terre" (the countries of the Earth) by the early encylopaedist Chevalier Barthélémy l'Anglais. Courtesy of Bibliothèque nationale de France.

volume). Additionally, *Recent Landscape Laboratories for Urban Futures* revives this tradition, where "woods go urban" (Nielsen et al. 2023). Over time, other programmes were also designed with an explicit relation to forests, particularly schools and sanitoriums – with the 1904 *Waldschule für kränkliche Kinder* (forest school for sickly children) in Charlottenburg (Berlin), by Walter Spickendorff, as well as Alvar Aalto's Paimio tuberculosis sanatorium (1928–1933) as canonical examples. The health benefits of forests for children are experiencing a revival of sorts with the global growth of contemporary 'forest schools' building on rich Scandinavian traditions of the 1970s (Dean 2019).

Forest Urbanism transcends the Garden City concept, even more radically recalibrating the way humans occupy the world, whereby settlements and agriculture are embedded within forests, liberated from reductive mono-functionality, amongst other things, through agro-forestry and urban forestry.

10. FORESTS, CITIES, COLONIAL DEVELOPMENT AND NATION-BUILDING

Colonial urbanism, by nature, was problematic and confrontational. On the one hand, its extractivist nature obliterated massive forests – as in wholesale clearing for agriculture and mining. On the other hand, scientific forestry led to massive tree planting – for materials needed for extractivism – and changed the nature of forests. In terms of settlement, neighbourhoods for the colonists were conceived as garden cities, and trees were an inherent component of 'civilizing' native lands. Imported species to the colonies were matched with exported, 'free'-moving species of 'exotics' to botanical gardens and arboretums (emblematic colonial institutions), private trophy gardens and, eventually, as internationally sought after decorative (street and park) trees. Trees were forcefully moved across ecological zones by Portuguese, Spanish, British, French, Japanese and other empires (Orlow 2018). Lingering evidence of

1963 saw the launch of the first tree-planting campaign in the city-state of Singapore by Prime Minister Lee Kuan Yew. The legacy of the highly influential campaign endures, along with an increasing number of similar initiatives worldwide.

such movements include, for example, Asian tree species lining European streets and the proliferation of Australian eucalyptus in Asia and the Americas. Ultimately, the rather wild and massive redistribution of species has led to the transformation of natural history into botanical sciences (Baber 2016). Since colonial times, debates have ensued regarding the appropriateness of native and non-native trees/species. It is a discourse that is as much ecological as it is social. It is noteworthy that in post-independence (or during apartheid in South Africa), botanical gardens intensively turned attention to indigenous species, as part of identity politics. After the 1947 partition of India, forests and urbanism were developed in tandem in the famous plans for new capitals, with Constantinos Doxiadis in Islamabad and Le Corbusier in Chandigarh. Green infrastructure was powerfully linked to nation-building (Rinaldi 2023). The concept was extended to the entire city-state of Singapore when then Prime Minister Lee Kuan Yew launched the first tree planting campaign in 1963. This project was followed by the Garden City programme in 1967 and the annual National Tree Planting Day in 1971. In many parts of Asia, a legacy of afforestation and street tree-planting was part and parcel of progressive eras of nation-building. It simultaneously had a 'reconstruction' dimension – implying that after the catastrophe of colonial rule, a previous order had to be reconstructed. Contrary to architecture, where an international modernism (and its promise of development) was embraced, the larger environment was often restored to highlight local context-embeddedness. Indigenous tree species were carefully selected for Chandigarh (where the vegetational and built structure are interwoven as warp and woof) and Singapore (where independence coincided with the initiation of tree planting campaigns) (Rinaldi 2023). In the above-mentioned Asian examples of postcolonial reconstruction (in other words, nation-building), planting was linked to the creation of new environments. Such 'reconstruction' was also evident in period after the American Civil War, which not coincidentally paralleled the initiation of a national park system (predominantly conserving large, monumental natural areas as national parks) (Diamant & Carr: 2022). "Anti-slavery activism, Civil War, and the remaking of the federal government gave rise to the American public park" (NPS 2022). In that respect, Frederick Law Olmsted's 1865 "Yosemite Report" is a key text, expressing "the aspirational vision of making great public parks keystone institutions of a renewed liberal democracy" (NPS 2022). What comes together in this keystone institution of recovered liberal democracy, is the access for all to nature, reconceived as a public domain and as a necessary complement to and compensation for the unnatural city (at the moment that those communities are acquiring a metropolitan scale, fueled by industrialization and migration).

Forest Urbanism can build on the notion that urban forestry is as much social as it is ecological. Since colonialism – and re-articulated strongly in the past decade – the indigenous/native and exotic/non-native discourse has shifted strongly towards favouring the former, as attention shifts from primarily economic or aesthetic concerns towards an ecological sensibility.

11. FORESTS AS PARK SYSTEMS: FROM NEIGHBOURHOODS TO THE REGIONAL SCALE

From the seventeenth century onwards, nature was fundamentally transformed by designers in Europe; promenades of trees were planted as urban parks and developed by the likes of John Nash in London, Jean-Charles Adolphe Alphand in Paris, Peter Josef Lenné in Berlin, and Holger Bolm in Stockholm (Da Costa Meyer 2013; Forrest & Konijnendijk 2005). As urbanization rates grew exponentially from the eighteenth century onwards, parks even more explicitly appeared as natural counter-figures and compensatory spaces, soon followed by urban forests (such as the Bois de Boulogne and Bois de Vincennes in Paris, or the Walden in Vienna, etc.). When the threshold transforming a city into a metropolis was crossed, these and other (new) elements, such as parkways and conservation

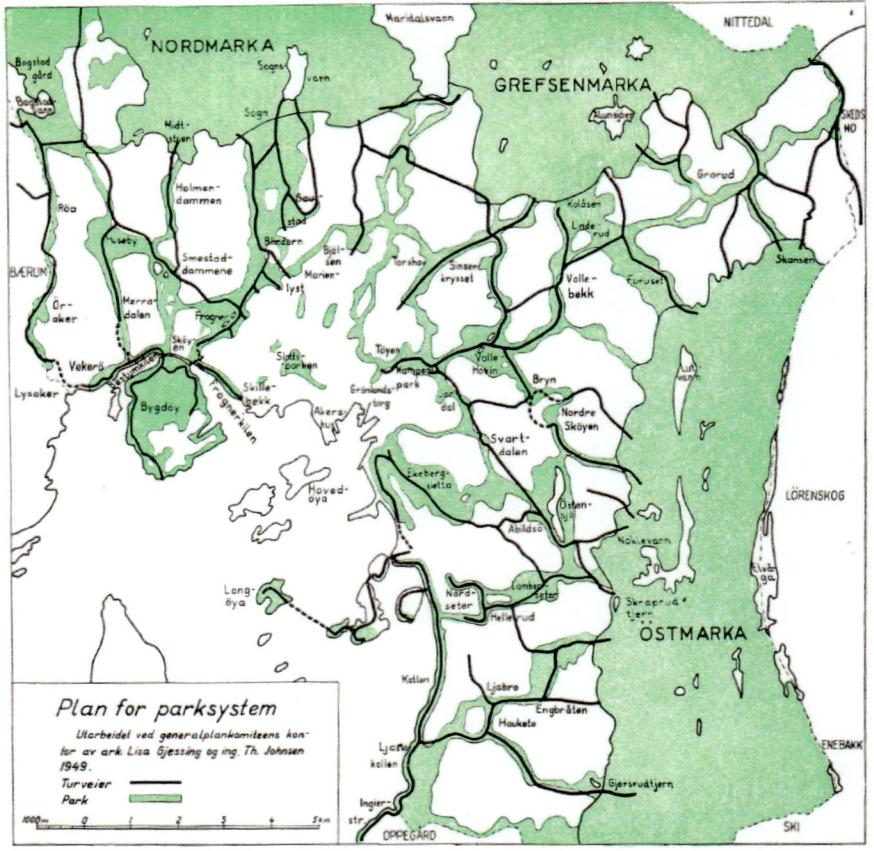

Within the 1950 General Plan for Oslo, Lisa Ijessing and Th. Lohnsø, designed a park system in which settlement and agriculture are embedded in the vast mountainous forests surrounding Oslo. Forest corridors create systematic accessibility through an extensive forest path network.

areas, were integrated into entire park systems, including the well-known plan of Jean Claude Nicolas Forestier (Alphand's protégé in Paris) for Buenos Aires (1902), in addition to his later plans for Havana (Forestier 1902). As put forward in Forestier's *Grandes villes et systèmes de parcs* (1908), park systems became an integral and necessary part of the wave of big master plans around the turn of the twentieth century, as developed, for instance, by Frederick Law Olmsted in Boston, New York, Baltimore, Chicago, and elsewhere. To some extent, all the large-scale forests, park and parkway plans, programmes and operations can be read as an important conceptual inversion. Instead of the urban environment, the landscape generated by extensive park systems – reaching beyond cities and including nature reserves and forests – was a framing structure into which the built landscape was embedded. Forestier's text could be considered foundational for the notion of *urbanisme paysager* (landscape urbanism) (Bonneau 2016: 63). The expansion of the imperial cities of Morocco in semi-arid conditions like Fez, Meknes, Rabat and Marrakech were defined by Forestier with their open space systems of gardens, parks and forest reserves, and so forth. Francois Ascher, French urbanist and sociologist, recognized in the 1990s the "return of landscape as the ordering principle of cities" (Ascher 1995: 239), while Sébastien Marot labeled such projects as an "alternative of the landscape" (Marot 1995). Earlier, Forestier phrased it a bit more modestly as landscape and parks being a complement, albeit absolute necessity. The metropoles in the making around the world understood, says Forestier, that "'the urban plan' is insufficient when not complemented with an overall programme and a special plan of the open spaces, inside and outside the city, at the moment and in the future" (Forestier 1905: 56) – in other words, by "a system of parks". A layer of a "park system" has become common practice in metropolitan master plans. Unfortunately, implementation is often neither quantitively nor qualitatively of the necessary scale and scope; park system concepts need a radical upgrade in light of the socio-ecological crises of the day.

Forest Urbanism stands for an urbanism in which the notion of (public) park systems are conceptually expanded in ecological terms, as well as in social and spatial terms, since the urban has long become a territorial reality. In Forest Urbanism, the urban is embedded within a forested landscape, and the forest structures the built environment.

12. FORESTRY AND REGIONAL PLANNING IN THE INDUSTRIAL AGE

In the mid-seventeenth century, British landscape architect and writer John Evelyn's *Sylva, or a Discourse of Forest Trees* was issued as a warning that continued growth of glassworks and iron industries would have dramatic consequences for British timber resources. Evelyn strongly advocated for an extensive reforestation programme and the systematic foundation of forests and parks in England. His work garnered significant attention since it asserted that reckless deforestation would compromise national defence, since the navy and merchant marine fleets were heavily dependent on wood. The book describes the various kinds of trees, their cultivation and the best use for each kind of England's timber (Evelyn 1664). At the end of the eighteenth century, Alexander von Humboldt, a founding father of ecology, as inspector of the Prussian Department of Mines, recognized the disastrous impact of then-nascent industrialization on ecology. His early nineteenth-century journeys to Latin American mines confirmed the devastating ecological effect and social devastation caused by colonial extractivism (Wulf 2016). Massive deforestation disrupted water cycles of river basins, and refining procedures permanently poisoned soil and water. Not surprisingly, colonial governments were soon forced to understand the necessity of afforestation in (mining regions) in order to safeguard water cycles and more generally heal the environment. A century later, massive forest reserves and afforestation programmes became standard practice to counterbalance and contain pollution from heavy industry and safeguard water resources of the Emscher River Basin – as evident in the Ruhr Region, the cradle of Germany's heavy industry. Belgian modernist Victor Bourgeois reconfigured the spatial development of the mining region of Charleroi by planning settlements not in the Sambre River Valley, but in the sloped forest areas beyond the windfall of polluted air (Bourgeois & De Cooman 1946). Forests quickly became a standard component in modern city models and regional development schemes – from the linear and socialist city concept of Miliutin

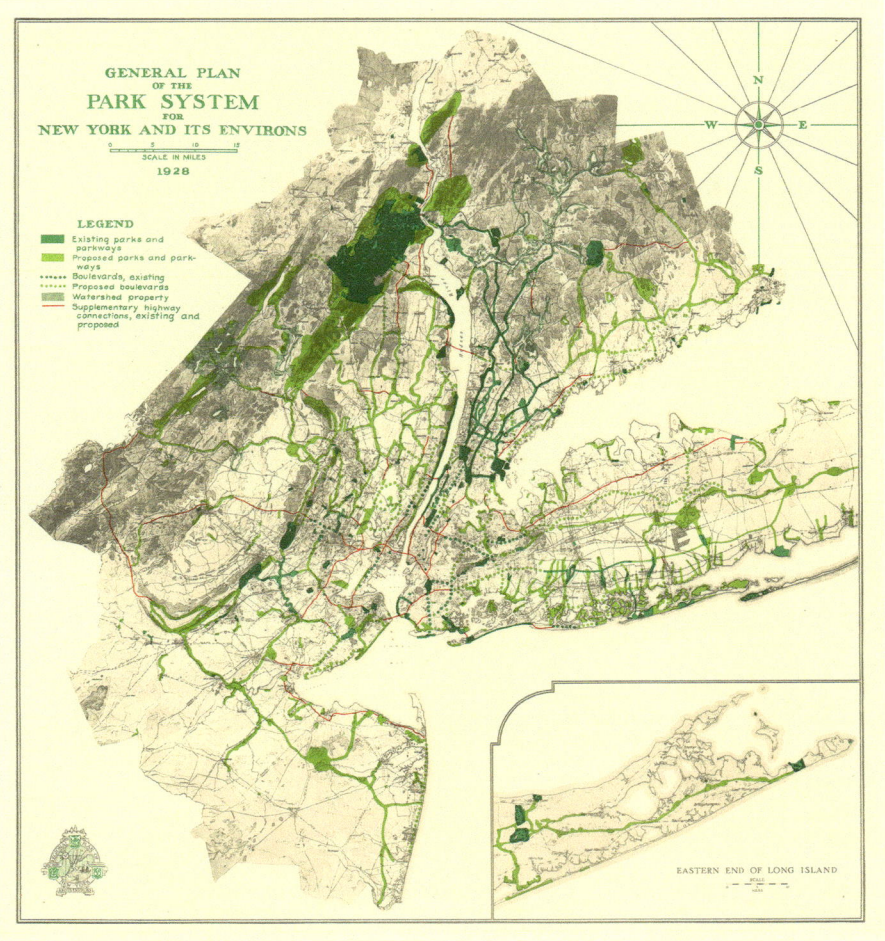

The "General Plan for the Park System of New York and Its Environs" (1928) is prototypical for plans in the golden age of industrialization, when visionary forest and park systems (including conservation areas) encompassed a regional scale. Courtesy of Regional Plan Association.

(Miliutin 1930) to Ludwig Hilbersheimer's "New Regional Pattern", where the occupation of the territory was split in workshops (industry), farms (agriculture) and settlements, all embedded within the forested landscape (Hilbersheimer 1949). The Finnish modern master Aalto (working with Otto-Iivari Meurman) was at the forefront of designing entire post-war districts embedded in forests, the most well-known of which is Tapiola New Town (1962–1963, 1967). Lafayette Park (1950–1960s) in Detroit, by architect Mies van der Rohe, urban planner Ludwig Hilbersheimer and landscape architect Alfred Caldwell is an American counterpart to the Finnish forest settlements. Lafayette Park was Charles Waldheim's primary precedent for "landscape urbanism" (Waldheim 2004). In regional planning, forests became complementary spaces where both conservation and planting gained momentum with industrialization, which in many contexts took place amidst other radical social, urban and political transformations (punctuated amongst others by wars and revolutions).

> **Forest Urbanism has river basins as their natural locales and scale. In (post-)industrialized river basins, new forests remediate pollution and repair natural water cycles, while settlements are oriented to suitable locations in the valley section, where the urban is embedded within a forested landscape.**

13. URBAN STREET TREES

For millennia, street trees have been systematically planned and planted on national scales and directly organized first by emperors, royalty and nobility and later by central, municipal and other governments. In China, *The Rituals of the Zhou Dynasty* (1100–770 BC) verifies that tree planting and maintenance by designated officials along city walls was obligatory. Initially, capital city streets and imperial highways were planted to provide separated royal passage, shelter against wind, provide shade, protect roads from flooding and perform specific visual functions. Whenever trees died, they had to be quickly replaced. "Tree plantings along city streets and country roads were considered as good moral behaviour

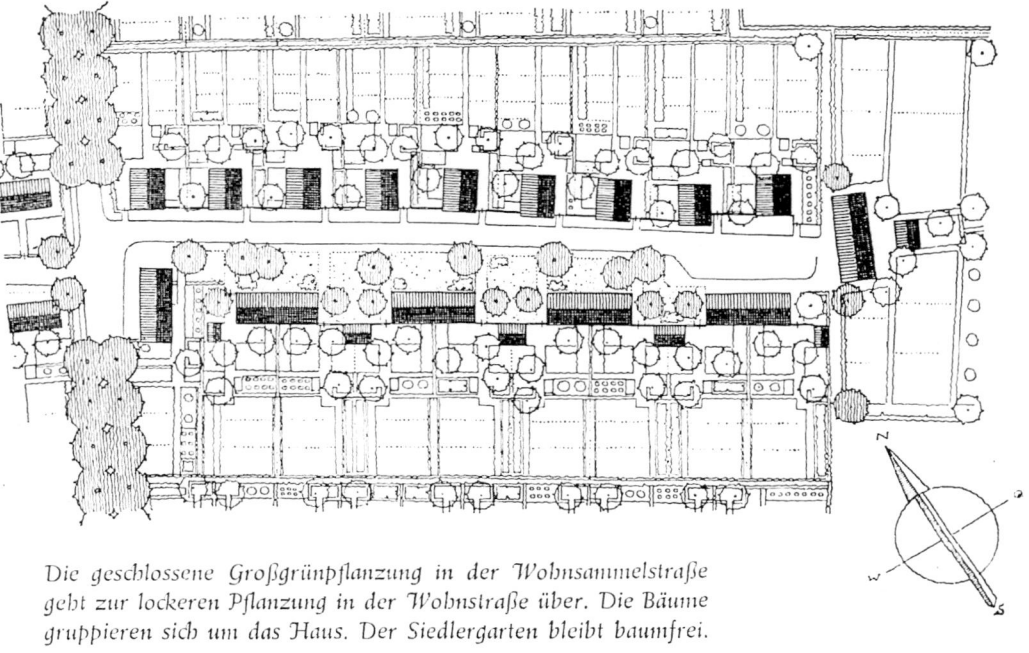

Die geschlossene Großgrünpflanzung in der Wohnsammelstraße geht zur lockeren Pflanzung in der Wohnstraße über. Die Bäume gruppieren sich um das Haus. Der Siedlergarten bleibt baumfrei.

Urban street and garden tree-planting schemes – in this case by H. Bronder for a prototypical German neighbourhood (1954) – are classic vegetal components of urbanism handbooks (existing since the eighteenth century).

and a blessing to the local people, and state officials were always memorialized for their contribution to the construction of greenways" (Yu et al. 2006: 230). Today, street trees are chosen for their urban suitability, resilience and aesthetics. The benefits of urban trees are widely acknowledged. Tree canopies cool urban heat islands and offer pockets of shade, absorb carbon through photosynthesis, produce oxygen, filter air pollutants, increase urban biodiversity by providing habitat for non-human species and even dampen noise (Forrest 2002). Additionally, landscape historian Sonia Dümpelmann underscores a deeper meaning of urban plants and trees and their capacity "to lift the human spirit, provide pleasure and psychological well-being and foster identity" (Dümpelmann 2022: 55). Her research underscores the importance of grassroots tree planting as part of wider social justice and civil rights activism in relation to environmental justice (Dümpelmann 2019). All that said, there are also negative consequences of urban street trees

– from both the perspective of the trees themselves and urban management. Large-scale urban tree planting only makes sense if followed by proper management and stewardship – which are both at least as important as planting, given that mature trees provide more benefits. The lifespan of street trees is significantly less than non-urban trees, and their overall health is compromised by constrained and often polluted soil conditions. Citizens appreciate urban street trees and benefit substantially from them, both in terms of physical and mental health. Nonetheless, too often this advantage does not translate into the necessary care and respect for urban street trees, which must adapt to the most unnatural conditions. From an urban management point of view, urban forests are seen as problematic due to pollen production, hydrocarbon emissions, green waste disposal, water consumption and displacement of native species by aggressive exotics (McPherson 1999: 42).

Forest Urbanism embraces the ancient tradition of street trees and seeks to further enhance their social relevance, while substantially improving their ecological conditions.

14. URBAN FORESTRY

In the Western world of environmental science, the term "urban forestry" was coined by Danish-Canadian forester Erik Jorgensen in 1965. More than three decades before the equally oxymoronic term "landscape urbanism," urban forestry was defined by Jorgensen as "the cultivation and management of trees for their present and potential contribution to the physiological, sociological and economic well-being of the urban society" (Jorgensen 1974, as quoted in Tree Canada). It was conceptualized to address urban trees

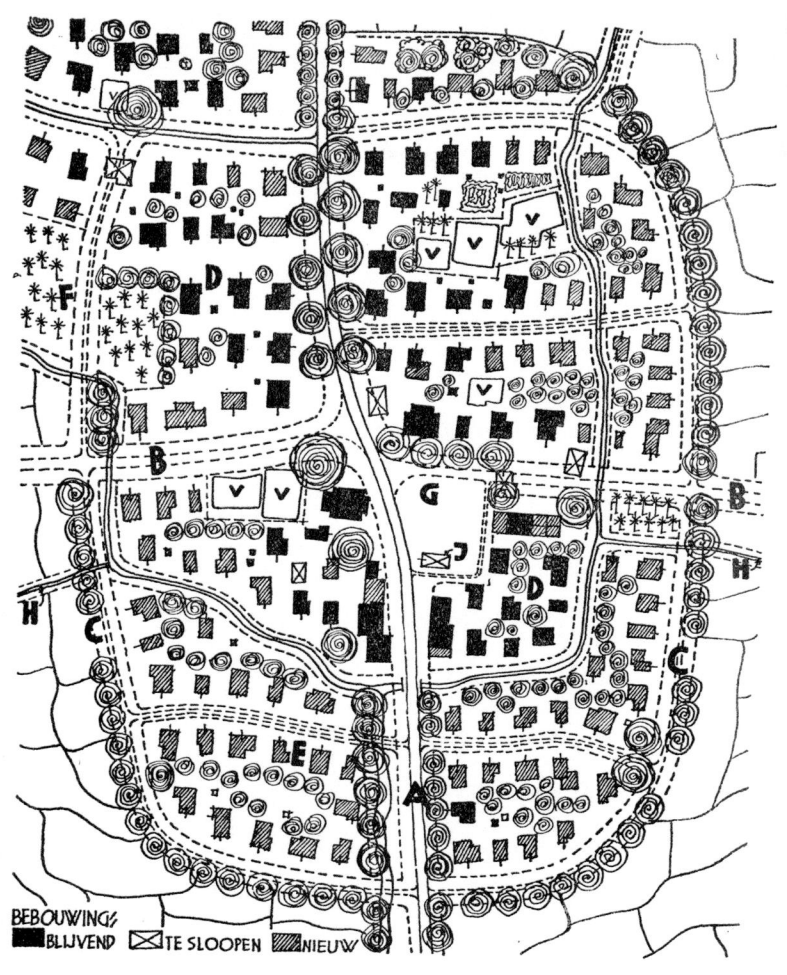

An ideal plan from 1947 for an Indonesian *kampung* (native township) by architect Charles Nix holistically integrates urban forestry and settlement planning.

beyond the single plant (street, shade and ornamental trees) towards an ecological community. It also confronted perceptually different realms: artificial versus natural, civilized versus wild, urban versus landscape. The explicit coupling of dichotomous notions (and the worlds they encompass) led to a scientific paradigm shift inasmuch as the ecological system created by urban trees is nowadays considered as a critical part the urban infrastructure (not so long ago still exclusively thought in hard engineering terms). Urban forestry and urbanism are both practical disciplines that immediately result in tangible changes. Relatively recently developed as a domain, urban forestry as an integral (sometimes implicit) part and parcel of urbanism has a long tradition, however (Konijnendijk 1997; Konijnendijk et al. 2005), induced by cultural practices involving traditions (what French town square and avenues do not have plane trees?). Caroline Mollie (amongst others, including Hugo Koch in the 1920s and Hans Bronder in the 1950s) convincingly demonstrated how trees have historically been the base component of a vegetal urbanism that is interwoven with the city (Koch 1921; Reichsheimstättenamt der DAF 1939; Bronder 1954; Mollie 2009). Urban forestry was also motivated by public health concerns as reforestation programmes during industrialization (as in the heavily polluted industrial Ruhr Region of the early twentieth century).

Forest Urbanism embraces the systemic contribution of urban forestry to the city, moving from considering trees in the city not as a collection, but as an ecological community. Forest Urbanism aims to elevate the role of forests to that of the structure of the city (rather than a mere element within the city). Once realized, urban forestry becomes Forest Urbanism.

15. URBAN FOREST ACCOUNTING

Since the advent of scientific forestry, there has been an increasingly technocratic approach to forestry and forests. Quantification became a necessity in relation to forest economies, and many nations and international institutions, such as the UN Food and Agricultural Organization (FAO), now include forest accounting

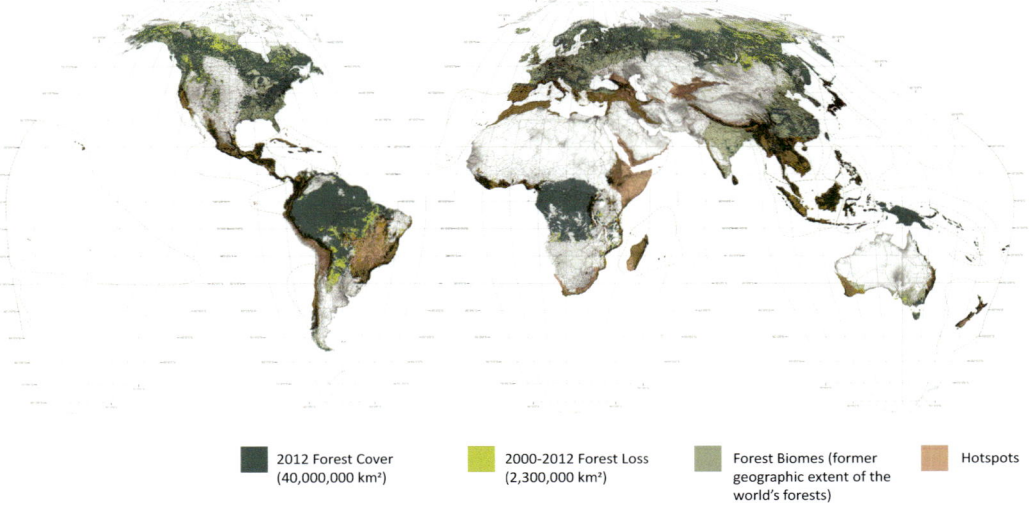

■ 2012 Forest Cover (40,000,000 km²) ■ 2000-2012 Forest Loss (2,300,000 km²) ■ Forest Biomes (former geographic extent of the world's forests) ■ Hotspots

Deforestation (forest cover, forest loss, forest biomes) in relation to the former geographic extent of the worlds' forests and hotspots, as mapped by Richard Weller, Claire Hoch and Chieh Huang in the *Atlas for the End of the World* (2015).

as standardized practices in relation to GDP (United Nations Millennial Ecosystem Assessment 2005). The notion of ecosystem services – of any positive benefit that ecosystems provide to humans – emerged as a term in the 1970s and has been refined by environmental and resource economists. It has gaining traction ever since, with a large boost following the 2005 United Nations Millennial Ecosystem Assessment (MEA). To date, there is a proliferation of checklists, evaluation criteria and toolboxes for forest accounting. In many urban contexts, urban forest accounting propelled by municipal budget cuts became a persuasive tool to justify open space funding and management. The felling, cutting and clearing (as well as selective harvesting) of forests is forever tabulated – although few statistics are wholly reliable and much evidence is prejudiced (Williams 2008: 346). The landscape architect, scholar, and practitioner Rosetta Elkin has revealed how trees have become "artefacts" and "units" in global climate-policy accountancy, standardized and commodified (including the notion of ecosystem services), where humanity has largely failed to appreciate the "aliveness" and agency of plants. She contends that "plant life" is misread, misunderstood, exploited and eradicated as trees are largely reduced to "plant material". Evidence of the urban tree accountancy euphoria is in the numerous million tree programmes that have been realized across the globe (including Los Angeles and Denver, 2006; New York City and Shanghai 2007; London 2011). While cities outdo each other with "million" tree programmes, countries make gigantic climate pledges with billions of trees. At the Davos Economic Forum of 2022, China made pubic an intention to plant 70 billion trees by 2030, while the US pledged to grow 995 million trees and the EU is committed to three billion trees through the European Green Deal (Campbell 2022). Nonetheless, these technocratic programmes encounter well-grounded scepticism (Elkin 2022). Quantitative logics urgently require qualitative complements and critical context responsive approaches.

Forest Urbanism makes pleas for a balance of quantitative and qualitative approaches in evaluating forests and their entanglement with the urbanized world. Shifting from the urban to national or continental scale, more thoughtful but no less systemic tree planting programmes are required.

16. BREAKTHROUGHS IN FOREST SCIENCE

Since the early 2000s, numerous scholars, including Italian botanist Stefano Mancuso, Canadian forest ecologist Suzanne Simard and Israeli plant scientist Tamir Klein, have made progress in plant behavioural sciences, scientifically documenting how plants are subjects (organisms) rather than objects. They have provided novel insights into the relationships of trees to soils, landscapes (including other trees) and humans. Collectively, their research focuses on the

Artist Katie Holten depicted subsurface forest networks comprised of roots and fungal mycelia that support interspecies communication and nutrient exchange, originally commissioned for and published in *Emergence Magazine* (2019-20).

communication and social life of trees and plant ecologies and their ability to adapt intelligently to changing environments. They have brought into focus the importance of subsurface forest networks comprised of roots and fungal mycelia which support interspecies communication and nutrient exchange. More specifically, Mancuso has been involved with the emerging field of plant neurobiology, seeking to understand how plants acquire and respond to information in the environment. Plant signalling, plant (root) intelligence and plant memory are part of a growing lexicon in forest science (Mancuso 2017, Mancuso 2020). Simard, in work with First Nations communities in British Colombia, focuses on the way different forest tree species exchange carbon through their interconnected mycelia. Her co-authored *Nature* article of 1997 was featured on the journal's cover with the title "Wood Wide Web", which has become a catch phrase in popular science, as has her notion of the "mother tree" (Simard 2021). Of importance is that this concept challenges the prevailing thinking that cooperation is of lesser importance than competition in ecology. Klein focuses on how trees cycle water and nutrients between leaves, stems and roots, quantifying their role in carbon budgets (Klein 2014). He focuses on the role of trees in global warming and, particularly, in adapting to drought. Ultimately, the developing understanding of the complex relationships between trees and other plants, along with the consideration of the agency of plants, is increasingly reflected in urban forestry decisions related to, for example, design, species selection, establishment practices and understanding and developing plant systems rather than individual trees. Trees fix soils, release nutrients and stabilize the climate.

Forest Urbanism works with the notion of a cooperative and complex 'wood wide web' in the urban context, where the non-human and human co-exist.

17. WILDFIRES AND THE WILDLAND URBAN INTERFACE (WUI)

Forests burn – naturally, accidentally and intentionally. On the one hand, disturbance ecology and unpredictable fire regimes are an essential component of forest dynamics and play a decisive role in maintaining mosaics of forest, grasses and shrubland landscapes. Early slash-and-burn approaches – for rotational cultivation and creating space for animal grazing – was (and remains) a common

Ecologies are disturbed in the wildland urban interface (WUI), and the area is obviously vulnerable to wildfires, as witnessed in Santa Rosa's Coffey Park, 90 kilometres north of San Francisco in Sonoma County, which lost thousands of homes and businesses in the 2017 Tubbs Fire.

practice of both subsistence and forest stewardship by Indigenous peoples across the world, who practiced a broad-spectrum agriculturalist-cum-forager system (Pyne 1982). Intentional burning was also used to facilitate hunting and promote the growth of desirable wild plants and herbs. On the other hand, contemporary forest fires are commonly viewed as destructive processes of transformation – to ever-encroaching urban, suburban and rural development in or near wildland vegetation. In the mid-1980s, the term wildland-urban interface (WUI) was coined; it is the area where human settlement intermingles with, or abuts, unoccupied wildland vegetation. The WUI is a focal area for human/environment conflicts, such as destruction by wildfires, habitat fragmentation, introduction of exotic species and biodiversity decline (Cohen 2008; Shannon & Kaufman 2018). Over time, the cyclic nature of the fire disturbance regime has shortened, while their unpredictability remains preeminent and there is a continued blurring of (sub)urbanism and forests. Such realities have led to human-centric wildland fire management, aggressive suppression and restrictive development codes. It is evident that it is challenging to revise the financial and real estate juggernaut of urbanization in fire-prone ecosystems. Nature, however, talks back (to humans) on its own (non-negotiable) terms. The exponential increase of scorched landscapes is potent evidence that humankind must profoundly rethink the nature and culture relationship. Stewardship (including living with fire and new land management notions, such as fighting fire with fire through controlled burning) must go hand in hand with rethinking settlement in high-risk areas (Struzik 2017).

Forest Urbanism has WUI as a primary frontline to, as evolutionary ecologist E.O. Wilson has pleaded, radically reconfigure humankind's relationship with the earth by defining areas that are explicitly only inhabited by non-human species.

18. REFORESTATION AND THE (IL)LOGIC OF AFFORESTATION

Deforestation has occurred in an ever-rising upward curve since humankind inhabited Earth. Often, reforestation and afforestation (tree planting in dryland biomes) are grouped together as necessary (quick) 'fixes' to deforestation. Without a doubt, there is a necessity to increase forest cover. The issue is how. Recently, reforestation in urban areas has gained popularity through the pocket Miyawaki forest movement, which is a process of creating (almost instantly) mini, multi-stratal quasi-natural forests that are very biodiverse and

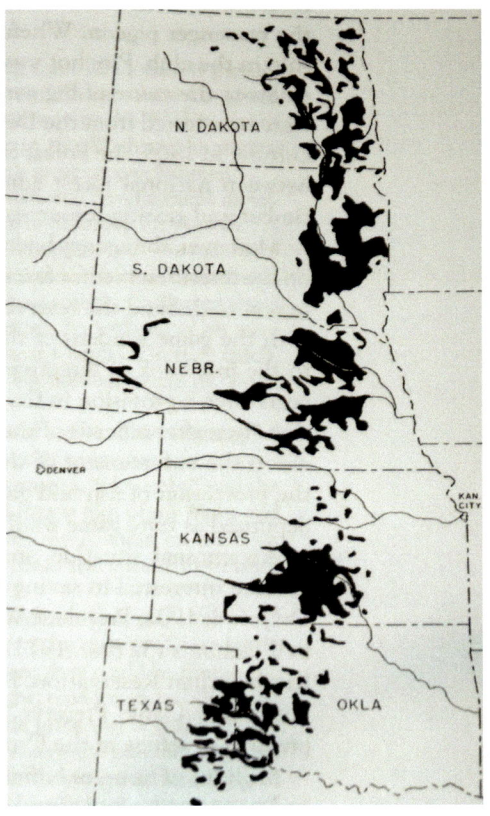 

The Shelterbelt Project (1933-42) of the US Forest Service, Civilian Conservation Corps and Works Public Administration planted millions of trees as walls to fix soils and provide protection against the Dust Bowl storms of the Great Plains.

allow humankind "to develop a sense of connection with the multitude of other species upon which we unknowingly depend" (Lewis 2022: 1). Reforestation evidently transcends the city. Afforestation relies on forestry and encompasses large-scale tree planning operations in otherwise treeless environments – including prairies, grasslands and deserts. The Prairie States Forestry Project (1934–42) was part of Franklin D. Roosevelt's New Deal projects. It planted a series of windbreaks (or shelterbelts) as tree walls from Texas to North Dakota to make life and agriculture in the Dust Bowl tolerable (Dahl 1940). The 7,775-kilometre-long and 15-kilometre-wide "Great Green Wall" across eleven Sahel countries in Africa is a contemporary attempt on a continental scale to increase arable land, reconstruct livelihoods and increase biodiversity and maintain native species (O'Connor & Ford 2014). Meanwhile, China is developing a number of enormous shelterbelt and dryland projects by the Chinese Forestry Administration, building on a long legacy linking back to Sun Yat-Sen and Mao Zedong (De Meulder & Shannon 2023–2). Past and present projects have met with varying degrees of success yet with significant costs and long-lasting environmental effects. Both reforestation and afforestation strategies come with critique. Foresters argue that protection of existing natural forests should take precedence over reforestation. Interestingly the French botanist and biologist Hallé, echoing Wilson's Half-Earth thesis (Wilson 2016), has proposed a radical project of human retreat to allow for the growth of a new primeval forest in western Europe (Hallé 2021). At the same time, it must be kept in mind that for centuries, the activity of people and most naturalists, botanists and the like have had, directly or indirectly, irreversible influence on the composition of primeval forests (Michon 2015: 29). Anthropologist Anna Tsing and historian and feminist Donna Haraway are amongst a number of scholars, demonstrating a way forward by proposing the "Plantationocene" as an alternative to the new era of the Anthropocene. For them, the current ecological crisis is rooted in logics of environmental modernization, homogeneity and control, which were developed on historical (often colonial) plantations (Tsing 2015; Haraway 2016). Radically reversing these logics is their message. Landscape architect Rosetta Elkin builds on the work of Tsing and Haraway and poignantly states in her recent book *Plant Life: The Entangled Politics of Afforestation* that, contrary to popular belief, tree planting is *not* inherently good. She warns of unintended negative consequences "as projects pair with world projects, casting tree planting as a solution to varying crises from soil erosion and air pollution to extreme drought and more recently global climate change. Because no one is asking any questions, tree planting appears to be an environmental obligation or, more outrageously, a solution. Tree planting does not moderate tree loss …" and "afforestation is an environmental venture that reveals an especially slippery slope because it represents progress in botany, forestry, and governance at the same time. The act of planting is not 'environmental;' it is entirely anthropogenic" (Elkin 2022: 2).

Forest Urbanism needs to be developed across scales, where it is most optimal to rewild (with humankind in retreat) and with new morphologies and typologies, so as to have meaningful co-existence of human and non-human species. Finally, massive deforestation has to stop and be replaced by sustainable stewardship. Indigenous Knowledge Systems and Practices (IKSP) deliver interesting indicators with regards to such stewardship.

19. TWENTY-FIRST CENTURY FOREST URBANISM: A NECESSARY PARADIGM SHIFT

The claim that the world is witnessing a planetary urbanism is an open door. There is hardly a place on earth without human settlement. There hardly exist settlements in pure isolation. Globalization has turned the planet in an "Earthopolis" (Nighthingale 2022). There is as well not a place on the earth where the environment is not affected. Indeed, as Rachel Carson explained, "Since the beginning of biological time, there has been the closest possible interdependence between the physical environment and the life it sustains (…) action and reaction between life and its surroundings have been going on ever since" (Carson 1963 as reprinted in Lear 1998: 230). Unfortunately, "it is true that, since the beginning of time, man has been a most untidy animal" (Carson 1963, as reprinted in Lear 1998: 228), dumping and releasing harmful substances on land, in the water and in the air. Pollution has become one of the

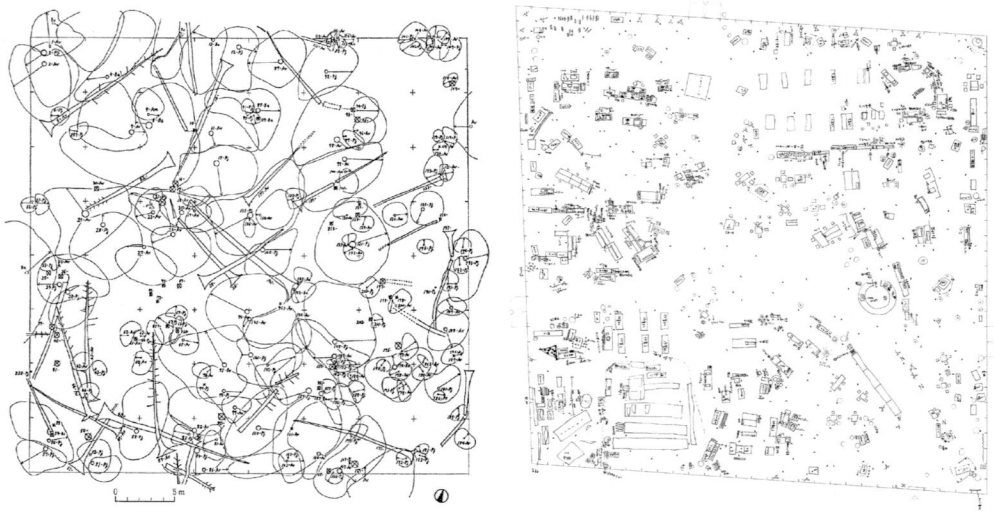

In *Another Scale of Architecture* (2019), Japanese architect Junya Ishigami painstakingly documents trees, their trunks and canopies (left) and notes how his office could occupy such a vegetal interior (right).

most vital problems of ecology and humankind. It is everywhere, in all forms and massive. Global warming is only starting to manifest tangible effects, fueling an ever more contentious social crisis. Current practices of urbanism – a discipline that originated as "a peaceful path to real reform", to paraphrase Ebenezer Howard – contribute more to the socio-ecological crisis than solve it. A paradigm shift is necessary. There need to be numerous initiatives of course-correction of the "most untidy animals" – man's behaviour. Thinking and understanding must be radically altered, and a myriad of changes needs to converge to realize the required essential transition. Contemporary transdisciplinary practices, such as forest urbanism, attempt to transcend the artificial distinctions and distance themselves from the artificial nature < > culture dichotomy that has been so deeply ingrained in disciplinary practices with a strong Western bias. New alliances with nature are required to reposition humans 'alongside other living beings.' There must be a radical reorientation between the dynamic action and reactions between living beings (man, cattle and plants of all kinds to begin with) and their physical environment. The earth's three primary land occupations – forestry, agriculture, urbanization – need to be hybridized and integrated to repair the damaging consequences of still-expanding mono-functionality. Fortunately, the dichotomy of ecology < > urbanism is being addressed in policies, plans and projects, and there is increasing discussion and understanding among foresters, land managers, landscape architects and urbanists, in dialogue with policy-makers and a wide range of relevant stakeholders, as well as in discussions across disciplines (Shannon et al. 2023). Vernacular agro-forestry around the world offers inspiring insights (Michon 2015).

Forest urbanism's challenge is operationalizing in concrete terms and tangible designs a redefined nature-culture relationship. It has to design the way in which humankind, along with other living beings, can concretely and responsibly occupy the world.

20. TWENTY-FIRST CENTURY FOREST URBANISM: NEW MORPHOLOGIES AND TYPOLOGIES

Since the urban is omnipresent on earth, forest urbanism of the twenty-first century must operate across scales and contexts – from the atmosphere to all kinds of environments on land (and sometimes on water). There is a clear mandate for more plants and forests in urban areas. *Forest Urbanism* goes beyond urban forestry and calls for the radical redefinition of settlement structures in relation to forests. Forest urbanism bridges landscape architecture and urbanism and reimagines land occupation to overcome the tripartite system of forestry, agriculture and urbanization through new hybrids and forms of multiplicity with regards to land occupation

The Mas Lombard Urban Development applies an emblematic forest urbanism figure, designed by Michel Desvigne (2020).

(De Meulder et al. 2019; Wambecq 2023). In practice, architects and landscape architects in France have conceived and developed projects that exemplify forest urbanism – including the Bordeaux Rive Droite (Right Bank) by Michel Desvigne Paysagiste (2005), Bois Habité (Inhabited Wood) by Agence Ter (2007) and the Grand Paris 2030 entry by Ateliers Jean Nouvel, Michel Cantal-Dupart and Jean-Marie Duthilleul (2008–2009). The need to develop urban forestry goes hand in hand with the ongoing need to restructure urban morphologies and to transform urban building typologies – to incorporate all the benefits of urban forestry, yet also to respond to new ways of structuring and inhabiting cities. One could think that the aesthetic interplay between architecture and forest – historically traced so eruditely by Gina Candell in her book *Tree Gardens* (Candell 2013) – has to urgently be complemented with case studies where aesthetics is replaced by ecological health. After all, in a long trajectory of meaning, trees have without a doubt a sacred, productive, and aesthetic meaning, but above and beyond their meaning is essentially biological and, by extension, ecological. In the meantime, several architects and urbanists – including Stefano Boeri and BIG – have already developed spectacular green buildings and urban districts. The Danish firm Effekt has developed more modest, though also more relevant visionary proposals for forest urbanism. The proposals also contain new housing typologies built exclusively with wood and designed to accommodate new family structures and forms of collective living embedded in forested environments. Academics and practitioners have similarly touted concepts like 'biophilic cities' and 'nature-based solutions'. Nevertheless, initiatives like these last ones run the risk of generic 'green-washing' unless, for example, issues of longer-term, sustainable management and stewardship are addressed. Scientists and designers alike need to engage with the evolving array of policy tools more vigorously, such as the transfer of (until now far too often mono-functional) development rights and land pooling. Multiplicity must (once again) become a guiding principle for territorial occupation and the co-presence, across scales, of nature (forests) and culture (agriculture and settlement) in order to shift current regimes of land management towards a 'commons' or to a public goods perspective, where urban forests and nature in the city can literally be afforded more space. Forests (in their diversity) – and as emblems of nature of which humankind is part – are structures in which settlements can be embedded as a promising reversal of the contemporary urbanization paradigm. It is important to learn the right things from Indigenous archetypical and proven sustainable forms of settling with/in forests and to re-envision socio-ecologically articulated settlement practices as they allow co-existence of humans and non-humans.

Forest Urbanism redefines ways of living that transcend the culture/nature distinction. It rearticulates stewardship of the environment, considering the intertwined ecological and social side of the challenge.

References

Abel, Mickey. (2017). *Medieval Urban Planning: The Monastery and Beyond*. Cambridge: Cambridge Scholars Publishing.

Albert, Bruce, & Kopenawa, Davi. (2022). *Yanomami, L'Esprit de la Forêt*. Arles: Actes Sud/Fondation Cartier pour l'art contemporain.

Anthwal, Ashish, Gupta, Nutan, Sharma, Archana, Anthwal, Smriti. (2010). Conserving Biodiversity through Traditional Beliefs in Sacred Groves in Uttarakhand Himalaya, India. *Resources Conservation and Recycling* 54, 11, 962–971.

Ascher, François. (1995). *Metapolis ou l'avenir des villes*. Paris: Odille Jacob.

Baber, Zaheer. (2016). The Plants of Empire: Botanic Gardens, Colonial Power and Botanical Knowledge. *Journal of Contemporary Asia*, 46 ,4, 659–679.

Bonneau, Emannuelle. (2016). *L'urbanisme paysager: une pédagogie de projet territorial. Architecture, aménagement de l'espace* (unpublished doctoral dissertation). Bordeaux, Université Michel de Montaigne – Bordeau III/Università degli Studi di Firenze.

Bourgeois, Victor, & De Cooman, René. (1946). *Charleroi, Terre d'urbanisme*. Brussels: Editions Art et Technique.

Bridel, J. B. (1798). *Manuel pratique du forestier. Ouvrage dans lequel on traite de l'estimation, exploitation, conservation, aménagement, repeuplement, des semis & plantations des forêts, avec les moyens de prévenir la disette des bois de construction & de chauffage*, Paris: Baudelot & Eberhart.

Bronder, Hans. (1954). *Grossgrüngestaltung und Städtebau*. Berlin: Deutscher Bauern Verlag.

Campbell, Charlie. (2022). China's Big Climate Pledge at Davos Sounds Promising, But Experts Are Skeptical. *Time*, 23 May 2023. https://time.com/6181214/china-tree-pledge-davos

Crandell, Gina. (2013). *Tree Gardens. Architecture and the Forest.* New York: Princeton Architectural Press.

Carrere, Ricardo, & Lohmann, Larry. (1996). *Pulping the South: Industrial tree plantations and the world paper economy*. London: Zed Books.

Carson, Rachel. (1963). The Pollution of Our Environment. Reprinted in Lear, Linda (ed.). (1998). *Woods. The discovered writings by Rachel Carson* (pp. 227–245). Boston: Beacon Press.

Cohen, Jack. (2008). The Wildland-urban Interface Fire Problem. *Forest History Today*, Fall, 20–26.

Chombart de Lauwe & Paul-Henri. (1964). Aspirations, images guides et transformations sociales. *Revue Français de Sociologie*, V, 180–194.

Dahl, Jerome. (1940). Progress and Development of the Prairie States Forestry Project. *Journal of Forestry*, 38 ,2, 301–306.

de Schlippe, Pierre. (1956). *Shifting Cultivation in Africa: The Zande System of Agriculture*. London: Routledge and Kegan Paul.

De Meulder, Bruno, & Shannon, Kelly. (2023-2). Towards and Asian Forest Urbanism. [Editorial in] *Landscape Architecture Frontiers*, 11, 1, 4–12.

De Meulder, Bruno, Shannon, Kelly, Nguyen, Minh Quang. (2019). Forest Urbanisms: Urban and Ecological Strategies and Tools for the Sonian Forest in Belgium. *Landscape Architecture Frontiers*, 7, 1, 18–33.

De Meulder, Bruno, & Shannon, Kelly. (2023). Topographies of Resistance and Resilience and Bathymetrical Realities and Dynamics of the Mekong and Saigon-Dong Nai Deltas. *Landscape Architecture Frontiers*, 11, 4, 10–27.

Dean, Stephanie. (2019). Seeing the Forest and the Trees: A Historical and Conceptual Look at Danish Forest Schools. *The International Journal of Early Childhood Environmental Education*, 6, 3, 53–63.

Descola, Philippe. (2015). *Par-delà nature et culture*. Paris: Gallimard, Collection Folio.

Descola, Philippe. (2014). All too human (still). A comment on Eduardo Kohn's *How forests think*. HAU: *Journal of Ethnographic Theory*, 4, 2. https://doi.org/10.14318/hau4.2.015.

Diamant, Rolf, & Carr, Ethan. (2022). *Olmsted and Yosemite. Civil War, Abolition and the National Park Idea*. Amherst (MA): Library of American Landscape History.

Di Palma, Vittoria. (2014). *Wasteland: A History*. New Haven and London: Yale University Press.

Didi-Huberman, Georges. (2023). *Tables de montage: Regarder, recueillir, raconteur*. Paris: IMEC.

Dümpelmann. Sonia. (2019). *Seeing Trees: A History of Street Trees in New York City and Berlin*. New Haven: Yale University Press.

Dümpelmann, Sonia. (2022). Plants. In Dümpelmann, Sonja (ed.), *The Landscape Project*. San Francisco: AR + D Publishing, 52-69.

Edwards, Penny. (2008). *Cambodge: The Cultivation of a Nation, 1860-1945*. Honolulu: University of Hawaii Press.

Escobar, Arturo. (2008). *Territories of Difference: Place, Movements, Life, Redes*. Durham: Duke University Press.

Elkin, Rosetta. (2022). *Plant Life: The Entangled Politics of Afforestation*. Minneapolis: University of Minnesota Press.

Evelyn, John. (1664). *Sylva, or A Discourse of Forest-Trees and the Propagation of Timber*. London: John Martyn for the Royal Society.

Forestier, Jean Claude Nicolas. (1902, reprint 1997). *Grandes villes et systèmes de parcs*. Paris: Norma.

Forrest, Mary. (2002). Trees in European cities – a historical review. In: Dunne, L. (ed.), *Biodiversity in the city,* 15–20.
Girot, Christophe. (2016). *The Course of Landscape Architecture: A History of our Designs on the Natural World, from Prehistory to the Present.* New York: Thames and Hudson.
Gregotti, Vittorio. (1996). *Inside Architecture.* Boston: MIT Press.
Grewe, Bernd-Stefan, & Hölzl, Richard. (2018). Forestry in Germany, c. 1550-2000. In Oosthoek, K. Jan, & Hölzl, Richard (eds.), *Managing Northern Europe's Forests: Histories from the Age of Improvement to the Age of Ecology.* New York and Oxford: Berghahn, 15-65.
Hallé, Francis. (2021). *Pour une forêt primaire en Europe de l'ouest: Manifeste.* Arles: Actes Sud.
Handel, Dan. (2011). First, the Forests: lecture at Canadian Centre for Architecture (CCA). https://www.cca.qc.ca/en/articles/issues/11/nature-reorganized/1500/first-the-forests. Accessed March, 2023.
Handel, Dan. (2017). It Goes on Like a Forest. In Springer, Anna-Sophie, & Turpin, Etienne (eds.), *The Word for World is Still Forest. Intercalations 4* (pp. 39–86). Berlin: Haus der Kulturen der Welt.
Haraway, Donna. (2016). *Staying with the Trouble: Making Kin in the Chthulucene.* Durham, NC: Duke University Press.
Hardin, Garrett. (1968). The Tragedy of the Commons. *Science, New Series* 162, 3859, 1243–1248.
Harrison, Robert Pogue. (1993). *Forests. The Shadow of Civilization.* Chicago: Chicago University Press.
Hilbersheimer, Ludwig. (1949). *The New Regional Pattern. Industries and Gardens, Workshops and Farms.* Chicago: Paul Theobald.
Howard, Ebenezer. (1898). *To-morrow: A Peaceful Path to Real Reform.* London: Swan Sonnenschein & Co.
Hughes, J. Donald, & Thirgood, J.V. (1982). Deforestation, Erosion, and Forest Management in Ancient Greece and Rome. *Journal of Forest History,* 26, 2, 60–75.
International Alliance of Indigenous-Tribal Peoples of the Tropical Forests. (1996). *Indigenous Peoples, Forest, and Biodiversity.* London: International Alliance of Indigenous-Tribal Peoples of the Tropical Forests & IWGIA.
Jones, Jill. (2016). *Urban forests. A natural history of trees and people in the American cityscape.* New York: Penguin Books.
Klein, Tamir. (2014). The variability of stomatal sensitivity to leaf water potential across tree species indicates a continuum between isohydric and anisohydric behaviours. *Functional Ecology,* 28, 6, 1313–1320.
Koch, Hugo. (1921). *Gartenkunst im Städtebau.* Berlin: Ernst Wasmuth.
Konijnendijk, Cecil. (1997). A Short History of Urban Forestry in Europe. *Journal of Arboriculture,* 23, 1, 31–39.
Konijnendijk, Cecil, Nilsson, Kjell, Randruo, Thomas, Schipperjin, Jasper. (2005). *Urban Forests and Trees: A Reference Book.* Berlin and Heidelberg: Springer-Verlag.
Lambert, Léopold (ed.). (2023). *Forest Struggles* [theme issue]. *The Funambulist Magazine. Politics of Space and Body,* 47.
Latour, Bruno. (2004). Why Has Critique Run out of Steam. From Matters of Fact to Matters of Concern. *Critical Inquiry,* 30, 225–248.
Laugier, Marc-Antoine. (1755, second edition). *Essai sur l'architecture* [Essay on Architecture]. Paris: Duchesne.
Le Guin, Ursula K. (1972). *The Word for World is Forest.* New York: Berkley Books.
Lear, Linda. (1998). *Rachel Carson: Witness for Nature.* New York: Henry Holt.
Lemaire, Tom. (2023). *Bomen en bossen. Bondgenoten voor een leefbare wereld.* Amsterdam: Ambo/Anthos.
Lewis, Hannah. (2022). *Mini-Forest Revolution. Using the Miyawaki Method to Rapidly Rewild the World.* Chelsea, VT: Chelsea Green Publishing.
Lowood, Henry E. (1990). The Calculating Forester: Quantification, Cameral Science, and the Emergence of Scientific Forestry Management in Germany. In Tore Frängsmyr, Helibron, J.L., Rider, Robin E. (eds.), *The Quantifying Spirit in the Eighteenth Century.* Berkeley: University of California Press, 315-343.
Mancuso, Stefano. (2017). *The Revolutionary Genius of Plants: A New Understanding of Plant Intelligence and Behavior.* New York: Atria Books.
Mancuso, Stefano. (2020). *Tree stories. How trees plant our world and connect our lives.* London: Profile Books.
Manning, Erin. (2023). *Out of the Clear.* Colchester: Minor Compositions.
Marsh, George Perkins. (1864). *Man and Nature. Or, Physical Geography as Modified by Human Action.* New York: Charles Scribner.
Marot, Sébastian. (2019). *Taking the Country's Side: Agriculture and Architecture.* Barcelona: Poligrafa Ediciones, S.A.
Marot, Sébastien. (1995). L'alternative du paysage. *Le Visiteur,* 1, 54–81.
McPherson, E. Gregory. (1992). Accounting for benefits and costs of urban greenspace. *Landscape and Urban Planning,* 22, 41–51.
Michon, Geneviève. (2015). *Agriculteurs à l'ombre des forêts du monde. Agroforestries vernaculaires.* Arles: Actes Sud & IRD Editions.
Miller, Char. (2004). *Gifford Pinchot and the Making of Modern Environmentalism.* Washington, DC: Island Press.
Mollie, Caroline. (2009). *Des arbres dans la ville. L'urbanisme vegetal.* Arles: Actes Sud & Cité Verte.

Moore, Glenn, & Atherton, Cassandra. (2020). The Veneration of Old Trees in Japan. *Arnoldia*, 77, I.4. https://arboretum.harvard.edu/stories/eternal-forests-the-veneration-of-old-trees-in-japan/. Accessed September 2023)

Mulder, Arjen. (2020). *The World According to Plants* [English translation of the chapter 'De plant als vriend']. Amsterdam: Uitgeverij De Arbeiderspers.

Miliutin, Nikolay Alexandrovich. (1974). *Sotsgorod: TheProblem of Building Socialist Cities* [reprint and translation of original edition, 1930]. Cambridge: MIT Press.

National Park Service (NPS). (2022). Olmsted and Yosemite: Civil War, Abolition, and the National Park Idea. Frederick Law Olmsted National Historic Site. https://www.nps.gov/articles/000/-olmsted-and-yosemite-civil-war-abolition-and-the-national-park-idea.htm. Accessed March, 2024.

Nielsen, Anders Busse, Diedrich, Lisa, Szanto, Catherine. (2023). *Woods go urban. Landscape laboratories in Scandinavia.* Wageningen: Blauwdruk Publishers and Swedish University of Agricultural Sciences (SLU).

Nightingale, Carl H. (2022). *Earthopolis. A Biography of Our Urban Planet.* Cambridge: Cambridge University Press.

O'Connor, David, & Ford, James. (2014) Increasing the Effectiveness of the "Great Green Wall" as an Adaptation to the Effects of Climate Change and Desertification in the Sahel. *Sustainability*, 6, 10, 7142–7154.

Orlow, Uriel. (2018). *Theatrum Botanicum*. Berlin: Sternberg Press.

Pearce, Fred. (2023). Sacred Groves: How the Spiritual Connection Helps Protect Nature. *Yale Environment 360*. https://e360.yale.edu/features/sacred-groves-religion-forests. Accessed August, 2023.

Pregill, Philip, & Volkman, Nancy. (1993). *Landscapes in History. Design and Planning in the Western Tradition.* New York: Van Nostrand Reinhold.

Puig de la Bellacasa, Maria. (2011) Matters of care in technoscience: Assembling neglected things. *Social Studies of Science*, 41, 1, 85–106.

Pyne, Stephen. (1982). *Fire in America: A Cultural History of Wildland and Rural Fire.* Seattle: University of Washington Press.

Reichsheimstättenamt der DAF. (1939). *Stadt und Landschaft.* Berlin: Verlag der DAF.

Rinaldi, Bianca Maria. (2023). Botanic nations: The aesthetic of the forest in Chandigarh and Singapore. *Journal of Landscape Architecture*, 1, 40–53.

Roberts, Patrick. (2021). *Jungle: How Tropical Forests Shaped the World—and Us.* New York: Viking Press.

Sarraut, Albert. (1923). *La mise en valeur des colonies françaises.* Paris: Payot.

Scott, James C. (2009). *The Art of Not being Governed: An Anarchist History of Upland Southeast Asia.* New Haven: Yale University Press.

Shannon, Kelly, Cavalieri, Chiara, Konijnendijk, Cecil. (2023) Urban forests, forest urbanisms and global warming: Developing greener, cooler and more resilient and adaptable cities. *Journal of Landscape Architecture, 44*, 8–13.

Shannon, Kelly, & Kaufman, Donielle. (2018). California is Burning: Rethinking the Wildland/ (Sub)urban Interface. Landezine. https://landezine.com/california-is-burning-rethinking-the-wildland-suburban-interface/. Accessed July, 2023.

Simard, Suzanne. (2021). *Finding the Mother Tree: Discovering Wisdom in the Forest.* New York: Knopf.

Springer, Anne-Sophie, & Turpin, Etienne. (2017). *The Word for World is Still Forest.* Berlin: HKW/K-Verlag.

Struzik, Edward. (2017). *Firestorm: How wildfire will shape our future.* Washington, DC: Island Press.

Tavares, Paolo. (2017). In the Forest Ruins. In Colomina, Beatriz, Hirsch, Nickolaus, Vidokle, Anton, Wigley, Mark (eds.). *Superhumanity: Design of the Self* (pp. 20–35). Minneapolis: The University of Minnesota Press.

Thomas, William L. (1956). *Man's Role in the Changing the Face of the Earth* [2 vols.]. Chicago, London: National Science Foundation and Wenner-Gren Foundation for Anthropological Research.

Thoreau, Henry David. (1854). *Walden, or life in the woods.* Boston: Ticknor and Fields.

Tree Canada. (2024). https://treecanada.ca/article/in-celebration-of-erik-jorgensen-the-inventor-of-urban-forestry/. Accessed March, 2024.

Trosper, Ronald, & Parrotta, John. (2012). Introduction: The Growing Importance of Traditional Forest-Related Knowledge. In *Traditional Forest-Related Knowledge: Sustaining Communities, Ecosystems and Biocultural Diversity*, 1–36. Dordrecht: Springer.

Tsing, Anna. (2015). *The Mushroom at the End of the World.* New York: Princeton University Press.

Turnbull, Colin. (1961). *The Forest People: A Study of the Pygmies of the Congo.* New York: Simon and Schuster.

United Nations Millennial Ecosystem Assessment. (2005). https://www.millenniumassessment.org/en/Index-2.html.

Vidalou, Jean-Baptiste. (2017). *Être Forêts. Habiter des territoires en lutte.* Paris: Editions La Découverte (Zones).

Waldheim, Charles (ed.) (2004). *Case: Lafayette Park Detroit.* New York: Prestel.

Wambecq, Wim. (2023). *Forest Urbanism: in the Dispersed Flemish Territory.* Barcelona: Fundacíon Arquia.

Ward, Stephen V. (1992). *The Garden City: Past, present and future*. Abingdon: Spon Press.
Wild, Rob, & McLeod, Christopher (eds.). (2008). *Sacred Natural Sites: Guidelines for Protected Area Managers.* International Union for Conservation of Nature (IUCN): Gland, Switzerland.
Williams, Michael. (2008). A New Look at Global Forest Histories of Land Clearing, *Annual Review of Environment and Resources,* 33, 345–367.
Wilson, Edward O. (2016). *Half-Earth: Our Planet's Fight for Life*. New York: Liveright.
Worster, David. (1993). *The Wealth of Nature. Environmental History and the Ecological Imagination*. Oxford: Oxford University Press.
Wulf, Andrea. (2016). *The Invention of Nature: Alexander von Humboldt's New World*. New York: Knopf.
Yu Kongjian, Li Dihua, and Li Nuyu. (2006). The evolution of greenways in China. *Landscape and Urban Planning,* 76, 223–239.

FORESTS & SCIENCE

I. THE 3 + 30 + 300 RULE FOR URBAN FORESTRY

Chiara Cavalieri,
Cecil Konijnendijk

The need for greening standards

Green spaces and trees contribute to better public health by enhancing mental well-being, encouraging people to be physically active, creating social meeting places and cooling. A recent study by ISGlobal (2022) looked at access to green space and vegetation cover in European cities, concluding that 60% of the urban population live in areas without sufficient green space. Increasing overall greenness could result in 43,000 less premature deaths annually. Decision-makers and planners are willing to integrate trees and green space in policies and programmes, but they call for clear, evidence-based guidelines and standards. Typically, guidelines and norms for green space provision have related to the overall greenness of a city, to the amount and share of public parks and green spaces of different categories and types and, more recently, to the provision of green space to all residents based on distance to the nearest green area. During recent years, urban tree canopy cover has become widely used an indicator of greenness and the provision of ecosystem services.

    The current ecological and climatic crisis has brought renewed attention to cities and territories and the need to make them greener. Future transformations should acknowledge contexts as products of both human and non-human operations, as well as the result of a deep intertwinement of deforestation and urbanization processes. It comes as no surprise that today the places where green spaces and trees are needed most are often places that used to have a very rich environment in terms of trees, forests and green infrastructure. Cities and territories are places built on unique environmental pre-existences (Rogers 1948), *milieu* with thick cultural and historical dimensions.

To further expand the concept of thickness, which should be carefully considered when developing guidelines for urban greening, there are three notions that rise to the fore. Firstly is the notion of *site-specificity*, the uniqueness of a place, with a specific climate, natural geography, quality of the soils and subsoils, proximity of water resources, and existence of areas naturally prone to forest environments. A second element is that of *time*. Urban landscapes have a long history, often revealed by the traces of a complex *palimpsest* (Corboz 1983), upon which society wrote – traced, erased and ultimately re-traced – new chapters of urban transformation (Cavalieri & Cogato 2020). The historical and cultural dimension – the *longue durée*, for example, of deforestation and afforestation over the centuries – co-exists with the cyclical temporality of seasons. Seasons that, according to different climates, bring additional meanings to urban forests, for example those of serving as storm management areas, moderating urban heat islands, or preventing erosion. The third notion is the one of *scale*. The transformation of urban landscapes is the result of different design operations in dialogue with various scales of actions and stakeholders, from the very small-scale possibly activated by communities, to medium- or large- and systemic-scale typically managed by planners and city-makers.

The 3 + 30 + 300 rule

An example of recent guidance for urban greening is the 3+30+300 rule. Proposed by Cecil Konijnendijk in 2021, it was developed in relation to the latest evidence of climate adaptation and public health benefits of trees and green space. The rule (or rule of thumb) is based on the need for humankind to see green, to live in green settings and to be able to use green space for recreational needs. In a recent paper (Konijnendijk 2022), these aspects are further expanded, with studies demonstrating the importance of visible green for one's mental well-being and ability to concentrate, of canopy cover in neighbourhoods for climate adaptation (primarily cooling) and a range of health benefits and of easily accessible parks and other public green spaces for physical, social and mental health.

Seeing trees at the plot scale, Atlanta, Georgia (USA). The view of trees from one's window has been found to improve mental well-being and the ability to concentrate and perform.

Robust tree canopy at the neighbourhood scale, Vancouver, British Columbia (Canada). A decent tree canopy cover in one's neighbourhood promotes health and contributes to cooling on hot days.

Recreational use of green spaces, Amsterdam, the Netherlands. Parks are important for health and wellbeing, and connections to other people.

The 3+30+300 rule includes three interrelated components.

3 trees visible from every home, school and place of work
Every resident in a city, town or even village should be able to see at least three trees from their home, school or place of work. These trees should ideally be well-established. Although the specific number "3", as such, is not supported by scientific evidence, research has shown the importance of visible green space, including trees and shrubs on mental health and well-being (see Konijnendijk 2022).

30% tree canopy cover in every neighbourhood
Based on current research, a 30% canopy cover should be a minimum at the neighbourhood level, and cities should strive for even higher canopy percentage. Every neighbourhood needs to be targeted, including new housing developments where there are opportunities to integrate trees from the onset.

300 metres from the nearest park or green space
In line with research and World Health Organization (WHO 2017) recommendations, every citizen should have a large public green space within 300 metres, approximately a five-minute walk or so, from their home. WHO suggests a public green space of at least one hectare to allow for a range of activities.

The 3 + 30 + 300 rule.

3 + 30 + 300 in urban design strategies

Since cities are thick products of multiple and contested processes, the application of the 3+30+300 rule is not an easy task. Therefore, the rule should be understood as an orienting rather than a normative tool, as a rule of thumb that can inform with clear figures the necessary ecological transition for today. Moreover, contextualizing the three-fold rule clarifies the necessity of different *scales* of interventions and categories of users and city-makers. The *3-trees* level refers to users, in addition to small-scale and *site-specific* transformations, while advocating for an isotropic distribution of trees. In this case, the tree can be considered architecture in itself (Leonardi & Stagi 1982). The *30%-canopy* element mainly addresses city planners and policy-makers, as a suggested assessment tool. It refers to a rather medium-scale, along with leverage, in terms of ecological health and climatic mitigation. Lastly, the *300-metres* component is an indication that touches the neighbourhood and city scales and therefore includes city planners, local communities and citizen-scientists. Such a normative regulation could be difficult to implement as such because of the complexity of zoning regulations, pre-existing property structures and their *history*. Nevertheless, the 3+30+300 rule can be a powerful orientation tool for situated design strategies that re-use, re-think and invent new typologies of green spaces, including those that work within the palimpsest nature of places, going beyond generic solutions and having the ability to combine environmental performance with other functions, both for human and non-human ecologies.

References

Cavalieri, C., & Cogato Lanza, E. (2020). Territories in Time: Mapping Palimpsest Horizons. *Urban Planning*, 5, 2, 94–98. https://doi.org/10.17645/up.v5i2.3385.

Corboz, A. (1983). Le territoire comme palimpseste. *Diogène*, *121*, 14–35.

ISGlobal. (2022). *ISGlobal Ranking of Cities*. https://isglobalranking.org/. Accessed October 5, 2023.

Konijnendijk, C. (2021). Promoting health and wellbeing through urban forests: introducing the 3-30-300 rule. *IUCN Urban Alliance blog*, February 22, 2022. https://iucnurbanalliance.org/promoting-health-and-wellbeing-through-urban-forests-introducing-the-3-30-300-rule/. Accessed October 1, 2023).

Konijnendijk, C. (2022). Evidence-based guidelines for greener, healthier, more resilient neighbourhoods: Introducing the 3-30-300 rule. *Journal of Forestry Research*, 1-10. https://link.springer.com/article/10.1007/s11676-022-01523-z. https://doi.org/10.1007/s11676-022-01523-z.

Leonardi, C., & Stagi, F. (1982). *L'architettura degli alberi* [The architecture of the trees]. Milan: Mazzotta.

Rogers, E.N. (1955). Le preesistenze ambientali e i temi pratici contemporanei [Environmental preexistences and contemporary practical issues]. *Casabella continuità*, 204, February-March, 195.

WHO. (2017). *Urban green spaces: A brief for action*. Copenhagen: World Health Organization, Regional Office for Europe. https://www.euro.who.int/en/health-topics/environment-and-health/urban-health/publications/2017/urban-green-spaces-a-brief-for-action-2017. Accessed October 1, 2021.

II. ON THE NEED OF LARGE OLD TREES TO KEEP CITIES YOUNG AND VIBRANT

Rik De Vreese,
Bart Muys

The rise of urban trees as nature-based solutions

Creating and maintaining an urban treescape is recognized as a nature-based solution that contributes to increased urban liveability and a more resilient and sustainable city. Following the Billion Tree Campaign launched by UNEP in 2006, and in line with the increasing public support for tree planting, there are a high number of pledges towards planting additional trees in and around urban areas. As a result of the increased public interest in climate and nature (European Commission, 2021), and since trees and vegetation add to the value of the real estate, there are more and more trees included in designs for new urban (re)development, also in the context of mitigating urban heat islands.

Almost all tree-related ecosystem services flourish with time (Nowak & Crane 2000). And although stem increment culminates at rather young age (between five and 80 years depending on the species), the wood accumulation and carbon sequestration in tree biomass may continue for many more decades. Several studies show that most regulating services – including urban cooling, particulate matter filtering and neutralization of other pollutants – are size dependent and increase exponentially with the tree diameter and height. Cultural services, such as contribution to natural and cultural heritage, reach their peak at old growth stage, at ages in the order of centuries.

As a matter of fact, veteran trees provide a large range of microhabitats for biodiversity (Muys et al. 2022). An inventory of the city birds of Barcelona revealed that older city trees provide not only nesting habitats to common birds like Common Blue Tit (*Cyanistes caeruleus*) and Common Starling (*Sturnus vulgaris*), but

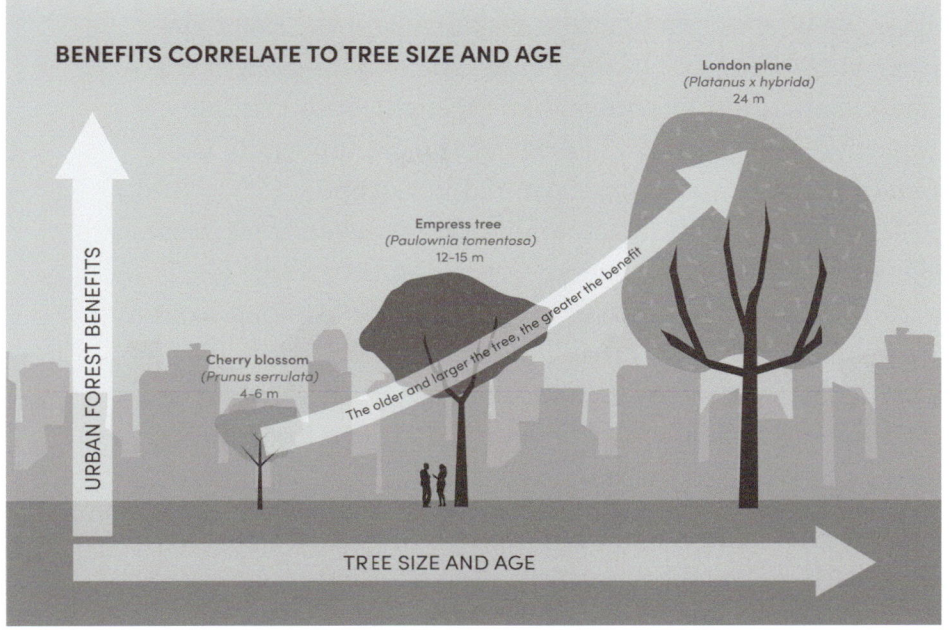

Benefits correlate to tree size and age.

Conservation of urban old-growth trees marking the limits of the Monte Testaccio, a fascinating ancient Roman amphora sherd dump site in Rome, Italy. Natural and cultural heritage are entangled here.

also for the more uncommon Spotless Starling (*Sturnus unicolor*) and Tawny Owl (*Strix aluco*) (Anton et al. 2017). Old-growth trees are also carriers of cultural heritage, like the eighteenth-century *Sophora japonica* at the Atrecht College of the old university in Leuven, one of the first honey trees planted in Europe.

In short, urban sustainability profits when trees grow old and big. Unfortunately, the average age of an urban tree is limited to 25 to 30 years (Roman & Scatena 2011). Looking from a greenhouse gas balance perspective, 25 years is merely sufficient to offset the carbon emissions for growing, transporting, planting and managing the tree (Kendall & McPherson 2012). The low average age of urban trees is due to several reasons:

1. trees planted are not appropriate for the site, or the site is not well prepared for tree planting (compacted, polluted or unsuitable soil; too limited root space);
2. trees are not well maintained (including watering during dry months);
3. trees cause damage to infrastructure (including pavements) or cause nuisances to people (falling leaves, blocking sunlight, negative impact on human health, personal injuries) and are cut to mitigate ill effects;
4. trees are in conflict with new developments, roads, and so on, and are cut to make space for development or redevelopment.

Many of such problems and conflicts can be avoided by advanced urban green planning, adapted species choice and improved information and education. All needs to start with the awareness that preserving existing trees is as important as planting more trees. The recognition of old growth is currently underrepresented in discourse and public actions. The focus on planting trees risks overshadowing the importance of preserving and managing existing urban trees in streets, parks, gardens and woodlands. Municipal budgets are under pressure, and the budget for tree management usually does not increase with the increasing number of trees. In other words, the average funding per tree typically decreases with tree planting. There are three recommendations to turn the tide.

Nuancing tree management

First, municipal budgets for tree management should increase and be in line with the number and size of trees managed. Funding for tree planting initiatives should not only include costs for the tree planting itself, but also for the follow-up management and maintenance for up to ten years, since maintenance during the establishment period can be labour- and cost-intensive. The funding for tree maintenance should be allocated in a specific fund that provides a yearly payment to the agency responsible for managing trees. Alternatively, as the maintenance of young trees is rather uncomplicated and safe, citizens can be engaged—for example, in watering trees, removing weeds and leaves (if needed) and performing limited pruning. The City of Essen (Germany) for example, runs the campaign "Watering Can Heroes" to engage citizens in watering trees in their neighbourhood.

Second, when planning new developments or redevelopments, existing trees should be preserved as much as possible.

Family active in the Watering Can Heroes campaign in Essen (Germany). They take stewardship of the tree in their street, watering it during dry periods.

Existing trees should be registered and monitored and should be considered as a guiding structure for development. As such, buildings and infrastructure could be planned in a way so that the impact on trees is as limited as possible. This protection should go beyond protecting trees from being cut. The buildings and construction interventions should also avoid impacting the root area and compacting the soil under the tree cover. When water drainage is temporally needed during construction works, precautions should be taken to avoid draining root space of the existing trees.

Finally, damage and nuisances can be avoided through planting "the right tree in the right place". Species selection should be well considered, taking into account the local conditions: not only the biotic and abiotic conditions should be assessed, but also the potential impact of the tree on the surrounding infrastructure, the allergenic potentials, and so forth. In light of climate change, the planner should also try to select future-proof trees that will withstand the future climate. In addition, in all species selections for the future, a mixture of tree species should be the guiding principle to enhance resilience of an old-growth urban treescape (Paquette et al. 2020).

Disclosure statement

Rik De Vreese is coordinating the CLEARING HOUSE project, funded by the European Union's Horizon 2020 Research and Innovation programme under grant agreement n°821242. The content of this document reflects only the authors' view. The European Commission is not responsible for any use that may be made of the information it contains.

References

Anton, M., Herrando, S., Garcia, D., Ferrer, X., Cebrian, R. (2017). *Atlas dels ocells nidificants de Barcelona*. Ajuntament de Barcelona. Barcelona: Edicions de la Universitat de Barcelona.

European Commission. (2021). *Special Eurobarometer 513 on Climate Change*. https://doi.org/10.2834/437.

Kendall, A., & McPherson, E.G. (2012). A life cycle greenhouse gas inventory of a tree production system. *International Journal Life Cycle Assess, 17*, 444–452.

Konijnendijk, C. (2022). Evidence-based guidelines for greener, healthier, more resilient neighbourhoods: Introducing the 3-30-300 rule. *Journal of Forestry Research*, 1–10. . https://link.springer.com/article/10.1007/s11676-022-01523-z. https://doi.org/10.1007/s11676-022-01523-z.

Muys, B., Angelstam, P., Bauhus, J., Bouriaud, L., Jactel, H., Kraigher, H., Müller, J., Pettorelli, N., Pötzelsberger, E., Primmer, E., and Svoboda, M., Van Meerbeek, K. (2022). Forest Biodiversity in Europe. *FSTP*, 13. Joensuu, Finland: European Forest Institute

Nowak, D.J., & Crane, D.E. (2000). The Urban Forest Effects (UFORE) Model: Quantifying urban forest structure and functions. In Hansen, M., & Burk, T. (eds.), USDA Forest Service General Technical Report NC-212 (pp. 714–720). St. Paul, MN: USDA Forest Service.

Paquette, A., Sousa-Silva, R., Maure, F., Cameron, E., Belluau, M., Messier, C. (2021). Praise for diversity: A functional approach to reduce risks in urban forests. *Urban Forestry & Urban Greening, 62*, 127–157.

Roman, L. A. & Scatena, F. N. (2011). Street tree survival rates: Meta-analysis of previous studies and application to a field survey in Philadelphia, PA, USA. *Urban Forestry & Urban Greening, 10*, 4, 269–274.

III. INTENTIONALLY INCLUSIVE URBAN FORESTRY

Colleen Murphy-Dunning

Perched along the Atlantic Coast between Boston and New York, New Haven, Connecticut, is a typical-sized U.S. city with a population of about 138,000 (2023), with 30% tree canopy across its 48.4 km². The city's residents reflect the racial diversity of the U.S. but reside in highly segregated neighbourhoods. One out of four New Haven residents lives below the poverty line, and across all races and ethnicities their poverty rates are substantially higher than state and national averages (U.S. Census Bureau, 2020). New Haven's neighbourhoods with majority persons of colour, lower income levels, and lower education unfortunately have significantly less canopy cover (Locke & Baine, 2014).

Although the United Nations Sustainable Development Goal 11 (target 11.7) only recently adopted fair and equitable access to urban green space for all as a global policy objective (UN 2022), New Haven's Urban Resources Initiative had long discerned the unfair distribution of trees in New Haven and was actively addressing the problem. Based on the evidence of the importance of trees' contribution to human health and well-being, Dr. Konijnendijk's 3+30+300 strategy could meet that U.N. target, that everyone should routinely see three trees, as well as live in an area with not less than 30%-canopy and be within 300-metres' distance of a public green space. To reach these goals, embracing an expansive view of the urban environment, as the field of public health does, would help, as New Haven was already attempting to do.

Public health professionals call the conditions where people live "social determinants of health" (SDOH) which include five key domains: education, health-care access, economic stability, social context and the environment. Taking a comprehensive view of natural resource management, akin to the public-health SDOH perspective, tree planting and stewardship can also be a tool for achieving social change. Foresters should create pathways for the public's personal involvement in decisions about their environment that directly affect their well-being. In New Haven, the Urban Resources Initiative, a nonprofit/university partnership based at the Yale School of the Environment, takes a fundamentally inclusive approach to managing New Haven's urban forest.

Urban Resources Initiative (URI) formalized its tree-planting efforts in New Haven with the launch of its community forestry programme ("Community Greenspace") in 1995. Volunteers participating in the Community Greenspace programme choose tree species and plant them in selected locations, such as along their street curb, with technical guidance and material support from URI. A narrow perspective of these community-led plantings might be limited to the ecosystem services provided, like the role of trees to provide cooling of homes. However, a more nuanced view – like the broader one followed by public health professionals – is enabling residents themselves to decide and take shared actions to improve their neighbourhood.

URI's inclusive approach is rooted in the belief that residents have insight about their community. Bridging local knowledge with forestry expertise from URI, residents can take effective action on what they identify is best for their environment. Opening that pathway of involvement leads to accomplishments that likely might not even occur to an urban forester. Of the thousands of volunteers who have planted trees with URI's help, their motivation to invest time and labour is usually for social, not environmental, reasons. The Greenspace volunteers often describe their purpose as: creating a safe place for children to play; making their neighbourhood beautiful; or remembering loved ones.

One powerful example is the recent founding of the New Haven Botanical Garden of Healing Dedicated to Victims of Gun Violence led by mothers who have lost their children. Guided by one mother's idea to dedicate a serene garden with a "tree of life" as a focal point for grieving families to recover in nature, the site accomplishes the parents' vision to create a sacred healing space for the hundreds of lives lost to gun violence in New Haven,

URI's GreenSkills adult team planting a street tree.

a plague that affects cities throughout the U.S. Supporting these parent volunteers has not only resulted in positive environmental outcomes through the planting, but more importantly has also brought a tranquil gathering space for families to connect and comfort each other.

By 2007, Greenspace volunteers had planted and nurtured over 1,600 trees in New Haven, with weekly watering yielding impressively high survival rates. Given their success, the local city government asked URI to lead all the city's tree-planting efforts. Tapping this opportunity, URI initiated its GreenSkills pilot, offering paid job training to high school students and, shortly thereafter, also to adults with obstacles to employment, such as a life history that included a felony conviction. Using the chance to plant trees to break down such employment barriers allows URI to link trees as a nature-based climate solution to the recidivism pattern borne by many individuals released from incarceration.

Reaching toward 10,000: Yale students celebrate with tree adopter after planting tree #9,955.

A 2022 U.S. Forest Service publication highlights the importance of integrating workforce development opportunities in urban forestry for those formerly incarcerated to address historic disinvestment in communities. Case studies in this report include URI's GreenSkills Program, as well as Pennsylvania's Correctional Conservation Collaborative, New Jersey's Tree Foundation Green Streets Program, Michigan's Line Clearance and Tree Trimming Program, and Greening of Detroit's Conservation Corps. URI is unique among the handful of existing urban forestry workforce programmes as it is based at a university.

In addition to developing a pathway of more inclusive urban forestry via paid job training for tree planting crews, URI's GreenSkills Rrogram works solely at the request of residents and businesses. This request-based procedure opens another avenue of involvement. When tree requesters, or "tree adopters", ask for a street tree, they can select the species they prefer from a palette of options that will work for that spot. In return for the free street tree planted by URI's GreenSkills crew, the tree adopter must commit to watering their new tree weekly for three years to ensure it successfully establishes. This process brings community involvement in the stewardship of the city's tree canopy by handing over the watering to the abutting property. This shared effort results in a markedly higher survival rate and an engaged public.

URI has now planted over 11,000 trees with Greenspace volunteers, with GreenSkills crews planting on behalf of over 3,000 tree adopters (as of November 2023). While mapping the location, species and survival rates of the trees to track their environmental impact remains essential, so does tracking public engagement. Of just over 175 adult apprentices who have learned about urban forestry operations, only 12% have returned to prison compared to the state average of over 50% recidivism. Yale forestry students have also mentored and taught over 200 high school students to identify tree species and help plant them. By collaborating with diverse partners in management activities and the community in decision-making, URI continues to strive to heighten both the biophysical and the invaluable social impact of trees.

References

Healthy People 2030, U.S. Department of Health and Human Services, Office of Disease Prevention and Health Promotion. Retrieved on October 20, 2022, from https://health.gov/healthypeople/objectives-and-data/social-determinants-health.

Konijnendijk, C. (2022). Evidence-based guidelines for greener, healthier, more resilient neighbourhoods: Introducing the 3-30-300 rule. *Journal of Forestry Research,* 1-10. https://link.springer.com/article/10.1007/s11676-022-01523-z. https://doi.org/10.1007/s11676-022-01523-z.

Locke, D. H., & Baine, G. (2014). The good, the bad, and the interested: How historical demographics explain present-day tree canopy, vacant lot and tree request spatial variability in New Haven, CT. *Urban Ecosystems, 18,* 2, 391–409.

United Nations. (2022). *Sustainable Development Goal 11.* Retrieved on October 2022 from https://sdgs.un.org/goals/goal11.

Zwerver, S. (2022). *Reducing Recidivism through Arboriculture Workforce Development. USDA Forest Service Eastern Region State and Private Forestry.* Retrieved January, 2024 from https://www.vibrantcitieslab.com/resources/reducing-recidivism-through-arboriculture-workforce-development/.

FOREST URBANISMS PROJECTS

I. LANDSCAPE AFTER NATURE
Trinity, La Défense, Paris (France)

Bureau Bas Smets
design implementation: 2011-2020   client: Unibail-Rodamco-Westfield

A new skyscraper has been built atop an existing urban boulevard in Paris La Défense. Surrounding the tower, a floating plaza connects the elevated esplanade of La Défense with the lower-lying urban fabric of streets and pavements. The man-made environment of the business district is surprisingly like that of a mountain range: glass facades, like glaciers, reflect the sun in all directions, while the artificial mineral hardscape resembles exposed rocks. The only missing component was a tree line that introduces a horizon among the peaks. The project was therefore developed with four distinct layers. The first layer considers the design as a mountain range, where every peak directly influences its surroundings. La Défense's skyline, like a mountain range, contrasts sharply with Haussmann's homogeneous nineteenth-century urbanism. In the latter, the street defines the disposition of the building; in the former, the addition of every tower is accompanied by a design for its adjacent public space. Another layer is the contour line along the slopes, above which no trees grow. If La Défense is understood as a mountain, it needs a tree line. The project therefore imagined a robust public space planted with many trees. The trees, suspended above the boulevard, create a new horizon and anchor the skyscraper in its environment. At the mountain base, the plaza itself is built above an urban boulevard, with a mere 52 centimetres of space between the finished level and supporting concrete structure. To achieve the tree line, the area for soil needed to be maximized. The result is a continuous fertile slab that allows roots to grow horizontally beneath the stone pavement. The final layer is a resulting mountain grove landscape of 50 trees that now grow atop the urban boulevard. Serving as a windbreak and providing shade, the trees create a microclimate between the tall towers. As in a natural grove, there is minimal to no undergrowth; the floor of the urban grove is a continuous stone pavement. Massive slabs of Lanhelin granite cover the plaza, the staircases, the elevators and the retaining walls. Like an avalanche, this mineral environment makes its way down towards the street level, enveloping everything in its path.

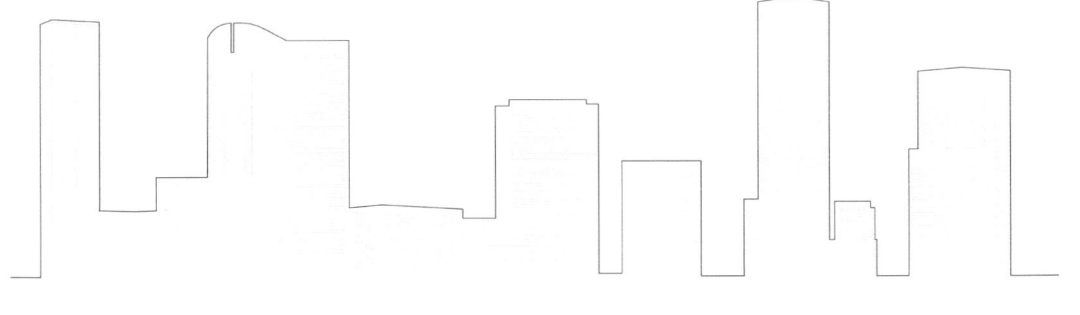

FOREST URBANISMS PROJECTS

FOREST URBANISMS

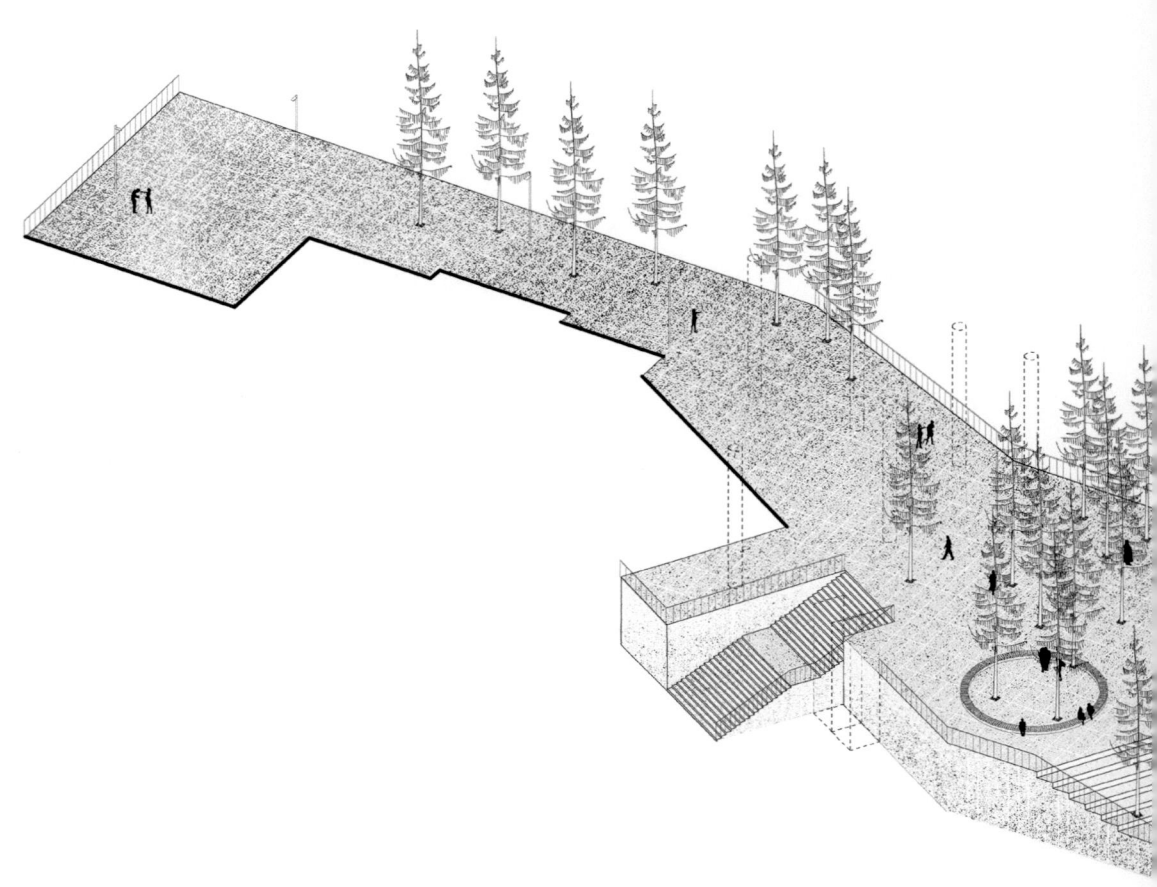

Trees have been strategically inserted into the concrete jungle of Paris La Défense, creating a pocket of forest urbanism that not only attenuates the climate, but also provides an expanded public realm in an increasingly privatized business district.

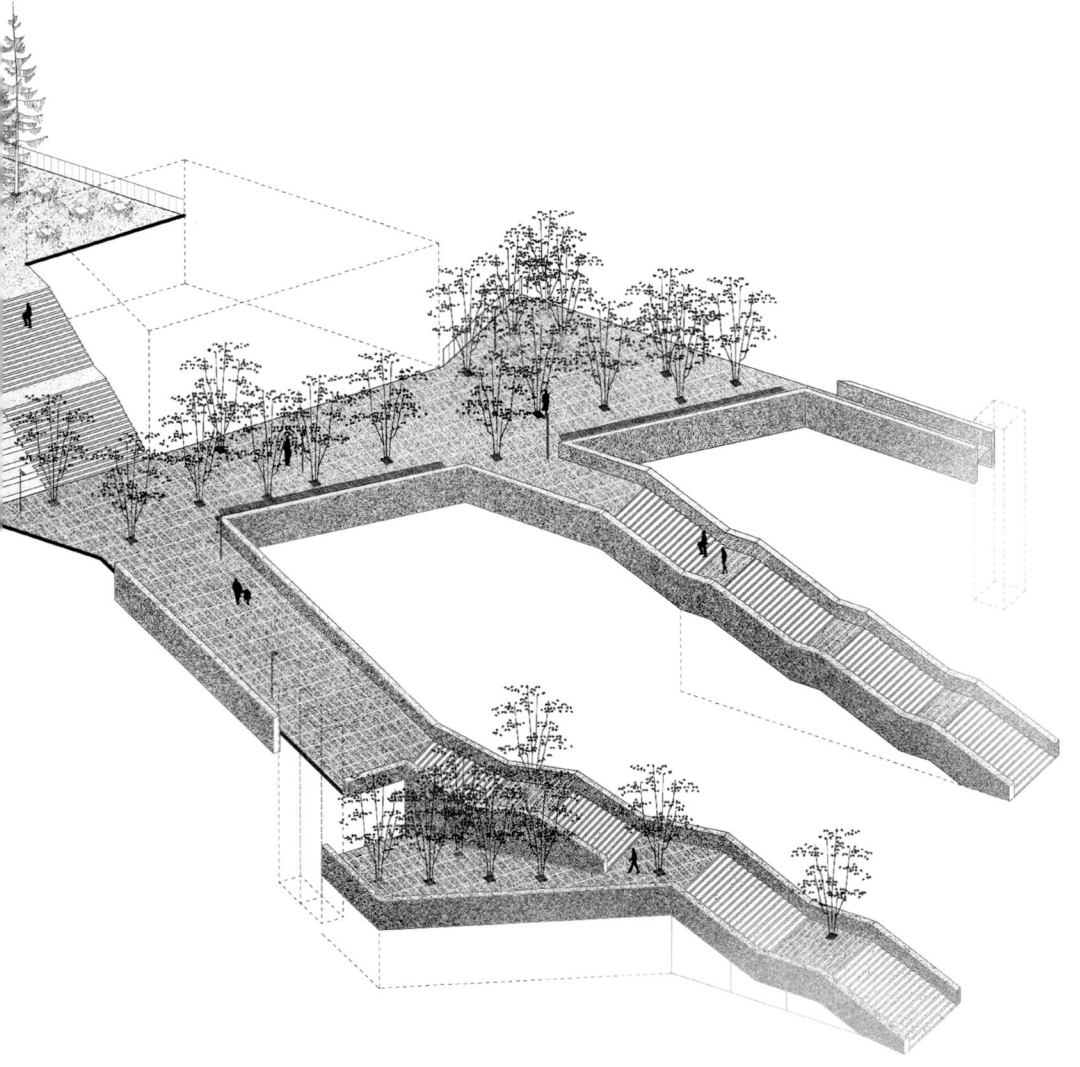

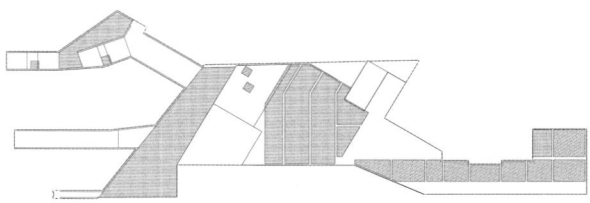

Fertile Slab

Alder trees were chosen for this project in two varieties: *Alnus glutinosa* and *Alnus glutinosa 'Imperialis'*. These trees can grow in shallow earth and can resist the use of salt in winter. The roots of these trees were prepared in the nursery to be compact and flattened. An automatic irrigation system is connected to a network of humidity detectors.

The trees grow where they can. Constraints from fire and universal access regulations mimic constraints found in a natural situation, like waterflows. There is no composition of trees, only opportunity for growth.

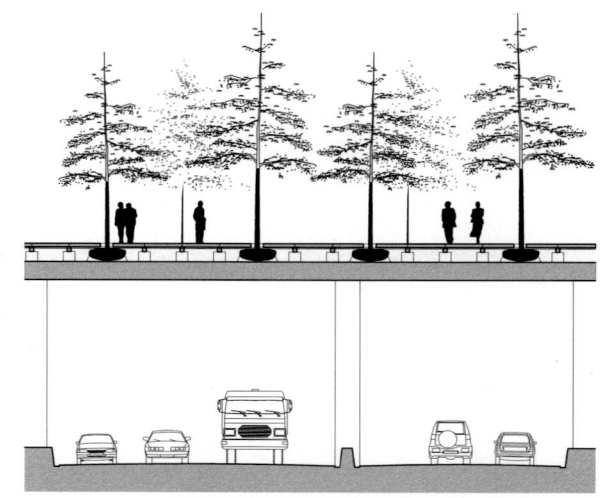

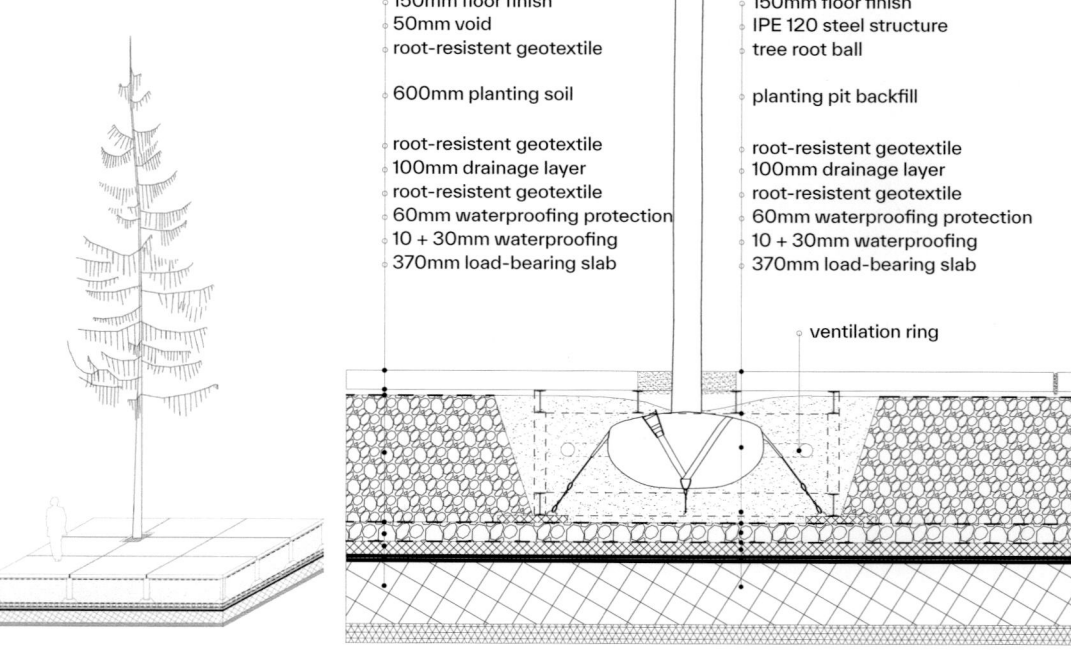

- 150mm floor finish
- 50mm void
- root-resistant geotextile

- 600mm planting soil

- root-resistant geotextile
- 100mm drainage layer
- root-resistant geotextile
- 60mm waterproofing protection
- 10 + 30mm waterproofing
- 370mm load-bearing slab

- 150mm floor finish
- IPE 120 steel structure
- tree root ball

- planting pit backfill

- root-resistant geotextile
- 100mm drainage layer
- root-resistant geotextile
- 60mm waterproofing protection
- 10 + 30mm waterproofing
- 370mm load-bearing slab

- ventilation ring

II. FISH TAIL PARK
Nanchang City, Jiangxi Province (China)

Kongjian Yu, Turenscape
with landscape architects Yu Hongqian,
Fang Yuan, Tong Hui and Jia Jianmin
design: 2017-2018; implementation: 2021

client: Nanchang Gaoxing Zhiye Property
Development Investment Co. Ltd.

Nanchang, the capital city of Jiangxi Province, is one of the largest cities on the middle reaches of the Yangzi River. Flooding and urban inundation during the monsoon season has been a chronic challenge. The problem has worsened in recent years due to the impacts of global warming and rapid development on lakes and wetlands, which has significantly reduced the landscape's water-regulating capacity. At the same time, surface water has been increasingly polluted by urban runoff, impacting the habitat of resident and migratory birds and other wildlife. New public spaces are being designed and built to fulfil the recreational needs of the growing population, but the biggest challenge – and opportunity – is how to develop integrated, efficient solutions to tackle the full spectrum of issues at a low cost and in a manner that can be replicable at a large scale.

The site is 51-hectare former fish farm reclaimed from a natural wetland. Roughly 30% of the site was a dump site for coal ash from power plants around the city. The surrounding area is slated for dense urban development, and the park is to be a catalyst for the development of a new district. The project developed a flood-adapted forest on the water. Inspired by the regional Poyang Lake's native monsoon-flood-adapted marsh landscape, the project includes tree species that are able to survive fluctuating water levels, including *Taxodium distichum, Taxodium distichum var. imbricatum* and *Metasequoia glyptostroboides*. The trees offer dense, verdant cover in the spring and summer and turn a vibrant orange in the autumn. Since fluctuating water levels often expose barren muddy shorelines, perennial and annual wetland plants were planted along the shorelines and island edges, and lotus plants provide highly efficient lake cover. Cut-and-fill of the dumped coal ash and soil from remnant dikes was utilized to create a lake with numerous islets. The lake can accommodate two metres of water-level rise, providing the capacity to catch a total 1 million cubic metres of stormwater inflow.

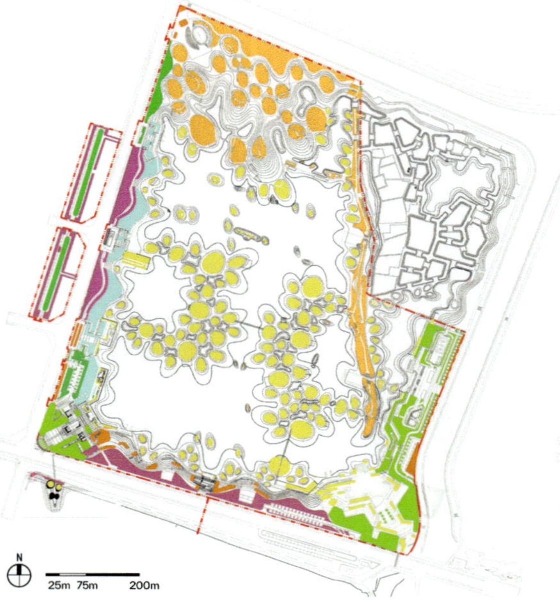
N
25m 75m 200m

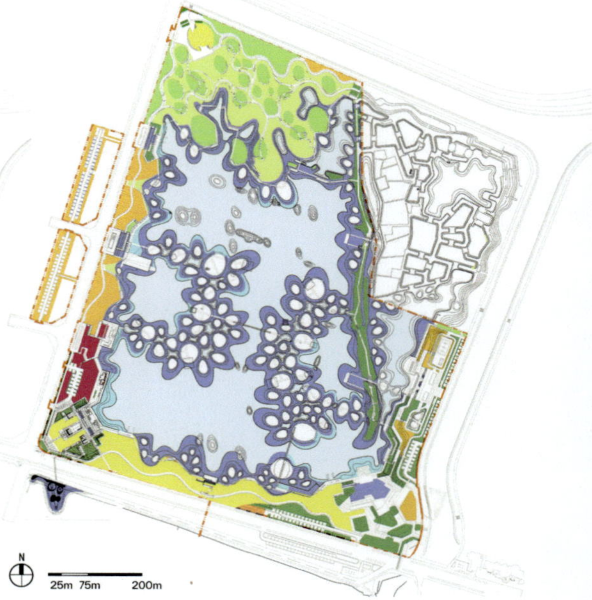
N
25m 75m 200m

- *Taxodium distichum var. imbricatum*
- *Sapium sebiferum*
- *Sapindus mukorossi*
- *Pterocarya stenoptera*
- *Ginkgo biloba*
- *Cinnamomum camphora*
- *Zelkova serrata*
- *Cynodon dactylon*
- *Reineckia carnea*
- Miscanthus
- Wild flower combination
- Combinations of grasses and flowers
- Fendai disordered seed grass
- Duchesnea indica
- Emergent plant
- Floating plant
- Underwater forest

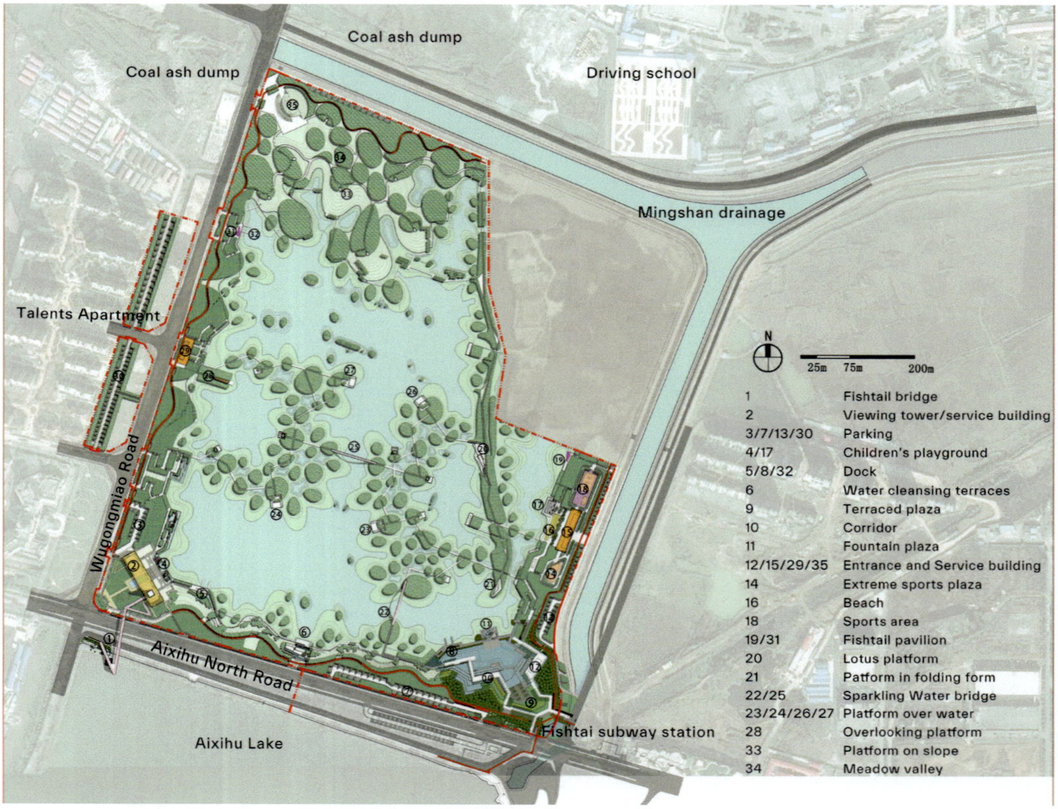
Coal ash dump
Coal ash dump
Driving school
Mingshan drainage
Talents Apartment
Wugongmiao Road
Aixihu North Road
Aixihu Lake
Fishtai subway station
N
25m 75m 200m
1 Fishtail bridge
2 Viewing tower/service building
3/7/13/30 Parking
4/17 Children's playground
5/8/32 Dock
6 Water cleansing terraces
9 Terraced plaza
10 Corridor
11 Fountain plaza
12/15/29/35 Entrance and Service building
14 Extreme sports plaza
16 Beach
18 Sports area
19/31 Fishtail pavilion
20 Lotus platform
21 Patform in folding form
22/25 Sparkling Water bridge
23/24/26/27 Platform over water
28 Overlooking platform
33 Platform on slope
34 Meadow valley

A network of low, flood-resistant boardwalks connects the islets, leading visitors through sheltered canopies to elevated concrete platforms nestled along the edge of select islets. Several of the platforms feature aluminium platforms and benches, affording visitors expansive views of the park and city beyond.

The floating forest heals an ecologically damaged urban landscape and operates as both a sponge and green lung for the bustling Chinese city of Nanchang. The forest is a foil to urban development, offering both respite for inhabitants and an oasis for biodiversity to thrive.

III. DEN HOUT 2040
Den Hout, Beerse (Belgium)

Wim Wambecq and Joris Moonen, MIDI

Design: 2017-2019

Client: PMV (Participatiemaatschappij Vlaanderen), Kerkfabriek Den Hout, Gemeente Beerse, Fam. Roefs

The project for 110 housing units and a new centre of Den Hout utilizes fragmented linear forests and tree lines (*houtkanten*, in Dutch), which were once part of the village's identity, as a structuring system for new urbanity. Instead of becoming filled by a generic allotment, the open space is strategically developed as an ecological park, bringing together the village's main functions from the canal to the main school. Two 'façades' frame a central space that alternates newly, more densely, built space with stretches of forest. Leftover linear forests are valorized and strengthened by allowing them to expand. Oriented east-west, thinner and thicker linear forest structures define a rhythm between the canal and the school. This 'canal forest' is a wide and lush urban forest that underscores the role of the canal in the village and the ecological and functional integration of Den Hout in the larger region. The forest is reinstated in the village centre through the design of densely forested urban collectives. Gradually, the monotonous allotment structures are turned into inhabited structures of linear forests, effectively creating an ecological continuum between the large patches of forest surrounding the village and the village centre itself. The forested character of "Den Hout" (literally translated: "the wood") is consolidated by an urban transition and absorbed into the forest matrix of the Campine, a region defined by the countless types of productive relations between urbanization and forests. Each act of urbanization can become an act of reforestation.

The existing urbanization in Den Hout, the gap in the future village centre and the project perimeter (in red).

A simulation of an imminent future if the current single-family urbanization continues.

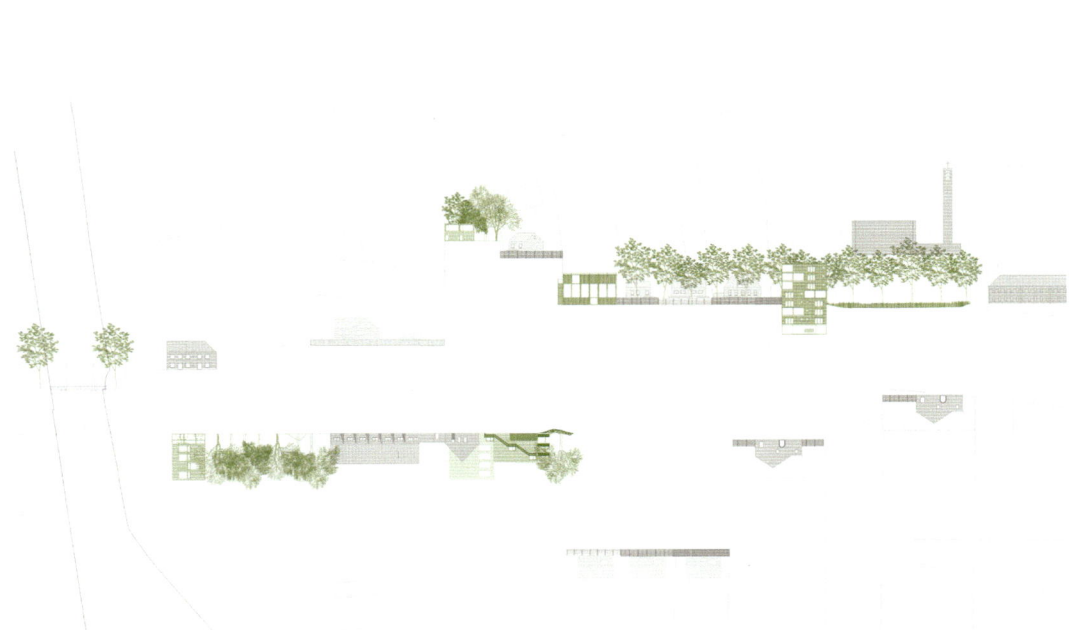

Village centre concept: a central space that brings together rhythms of forest and new urban forms.

The project designs an alternative to current practices that either hierarchize urban development above ecology and forest or demonize urban development for its destructive impacts on ecology. Forest creation and urban development become equal and mutually enriching terms – inseparable entities of one and the same forest urbanism habitat.

1775: "dries" + linear forest structure

1885: landscape inversion

1945-2000: hollowing out of forest

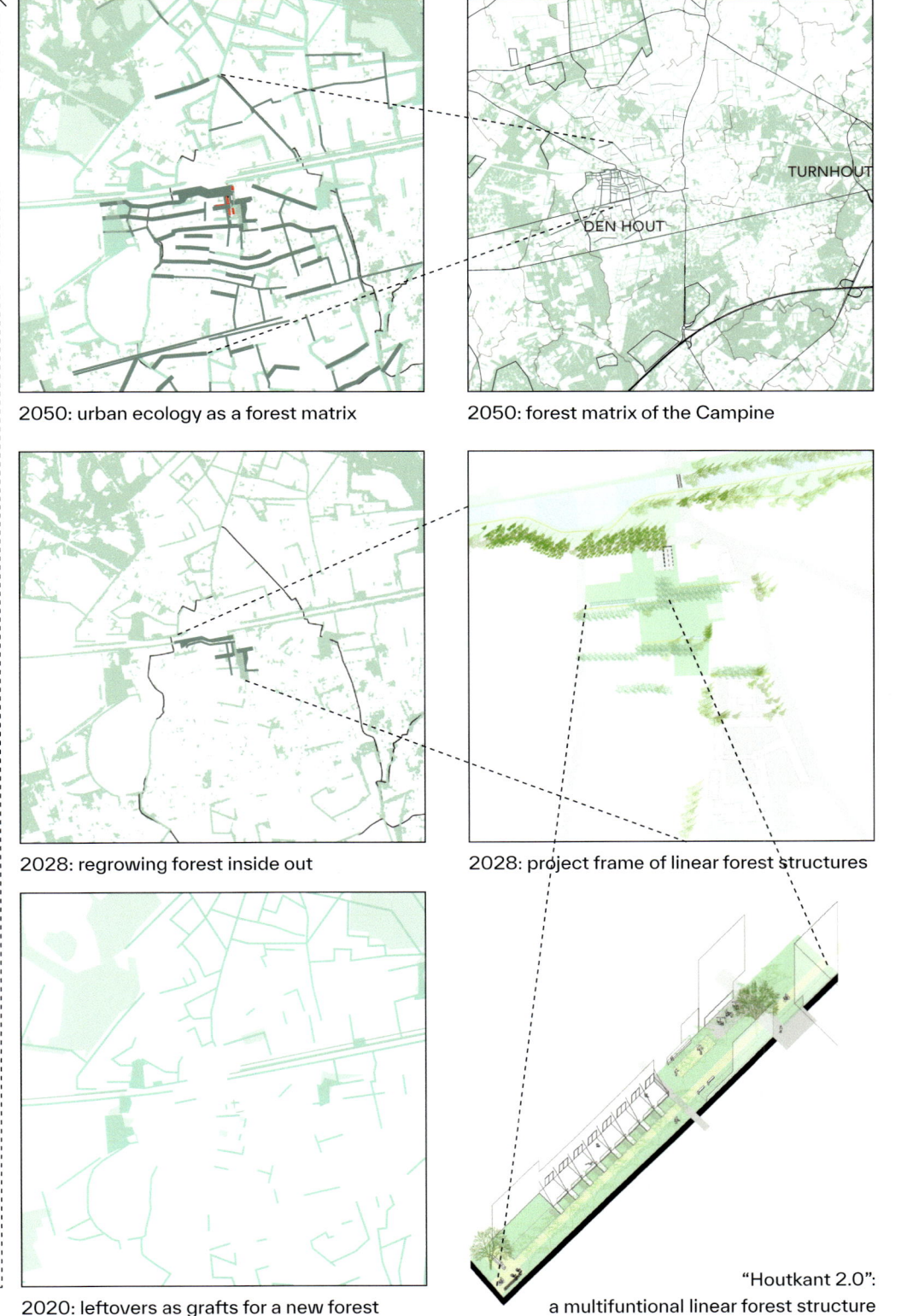

2050: urban ecology as a forest matrix

2050: forest matrix of the Campine

2028: regrowing forest inside out

2028: project frame of linear forest structures

2020: leftovers as grafts for a new forest

"Houtkant 2.0": a multifuntional linear forest structure

FOREST URBANISMS PROJECTS

FOREST URBANISMS

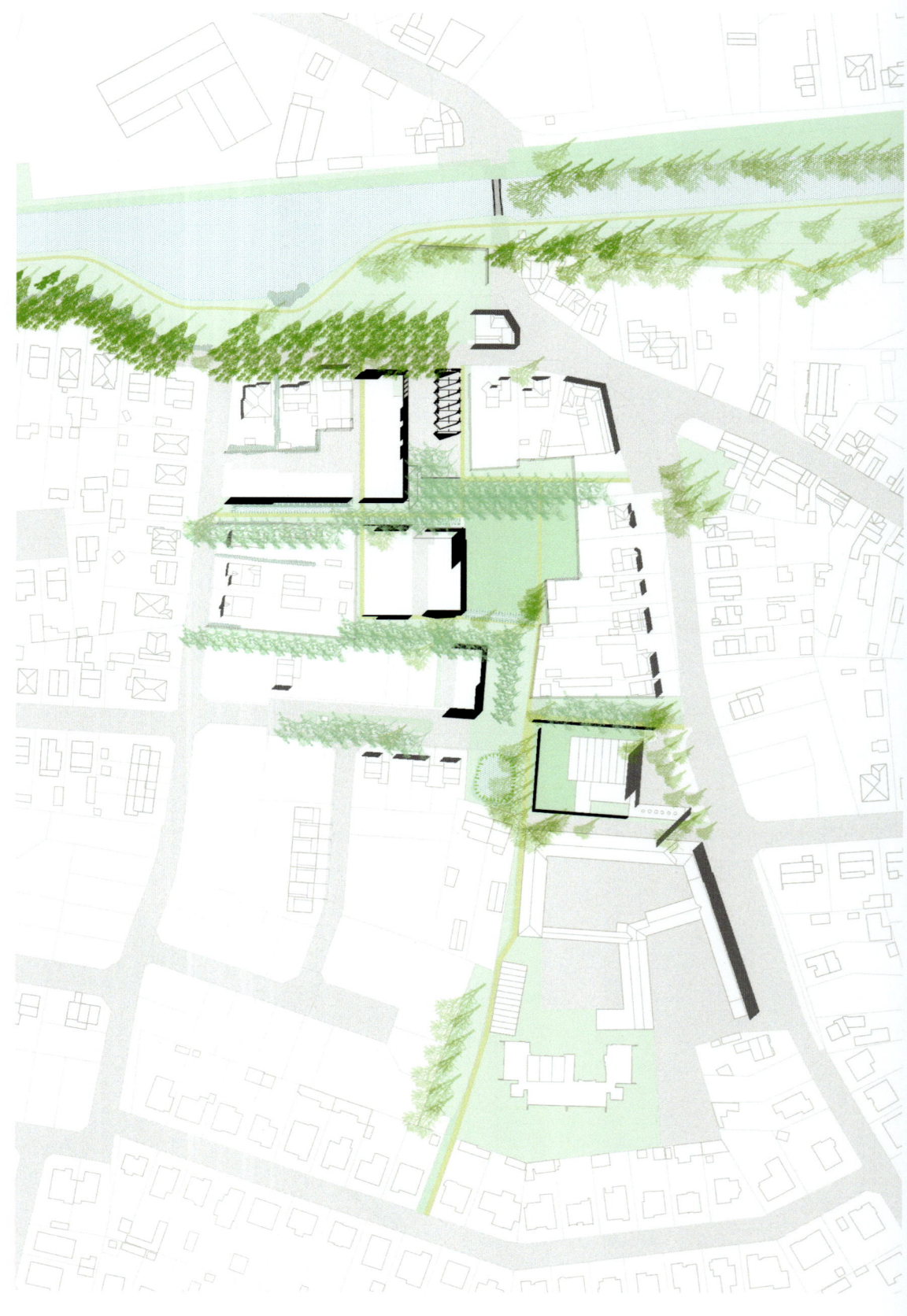

IV. SALVADOR ETHNOBOTANICAL GARDEN
Salvador (Brazil)

Embyá Paisagens & Ecossistemas
Design: 2019 (yet to be realized)   Client: Salvador City Council

Salvador, a Brazilian municipality with a population of three million inhabitants, is deeply rooted in historical and cultural connections to the African diaspora. The ethnobotanical garden is located in an existing urban fragment of the Atlantic Forest which hosts flora ranging from endemic and endangered to non-native specimens introduced through transatlantic trade. The design offers a comprehensive understanding of the rich plant knowledge embedded in Afro-Brazilian Candomblé rituals. Beyond its educational role, the garden seeks to create improved conditions for the growth, harvest and sacred use of specific plants. It also facilitates the production of seedlings for new plantings distributed throughout the metropolitan network of the city's *terreiros* (temples). Consequently, the project transforms the ethnobotanical garden into the focal point for the network of Afro-Brazilian religious sites, providing seedlings and hosting weekly sacred rituals. An ethnobotanical exhibition materializes in an open circuit of small and medium-sized gardens designed under the guidance of specific *Orixás*. Rather than adopting a mimetic or literal approach, the presence of the *Orixás* is intended to be sensed in land installations where visitors can perceive the existence of more-than-human beings interconnected with nature and its phenomena. Elements like the fragrance of particular plants, the colours of flowers, and the textures of leaves in intimate spaces collectively contribute to a diverse multisensory experience. The delineation of the *Orixás* garden areas was guided by the site's features, including water sources, natural light levels, slopes, prevailing winds and pre-existing plants.

(top) *Oxossi* spends most of his time alone in the forest. In his garden, a narrow trail leads to a single bench where he can sit and contemplate the surrounding nature in silence. (bottom) "*Ogum* cuts the lianas, cleans the tracks, takes care of the ways of the world" (Roger Bastide). The *Ogum* garden is crossed by a rock path that does a clearing in the woods. Laid down around the rock path, large trunks refer to cut branches and delimit new plantings linked to the *Orixá*.

The project re-articulates an open space as an ethnobotanical garden that underscores the sociocultural meaning of the tropical forest in the Afro-Brazilian context. The urban forest is new type of public space, linking to a larger cultural network by functioning as a nursery, a setting for rituals and a didactic landscape for visitors.

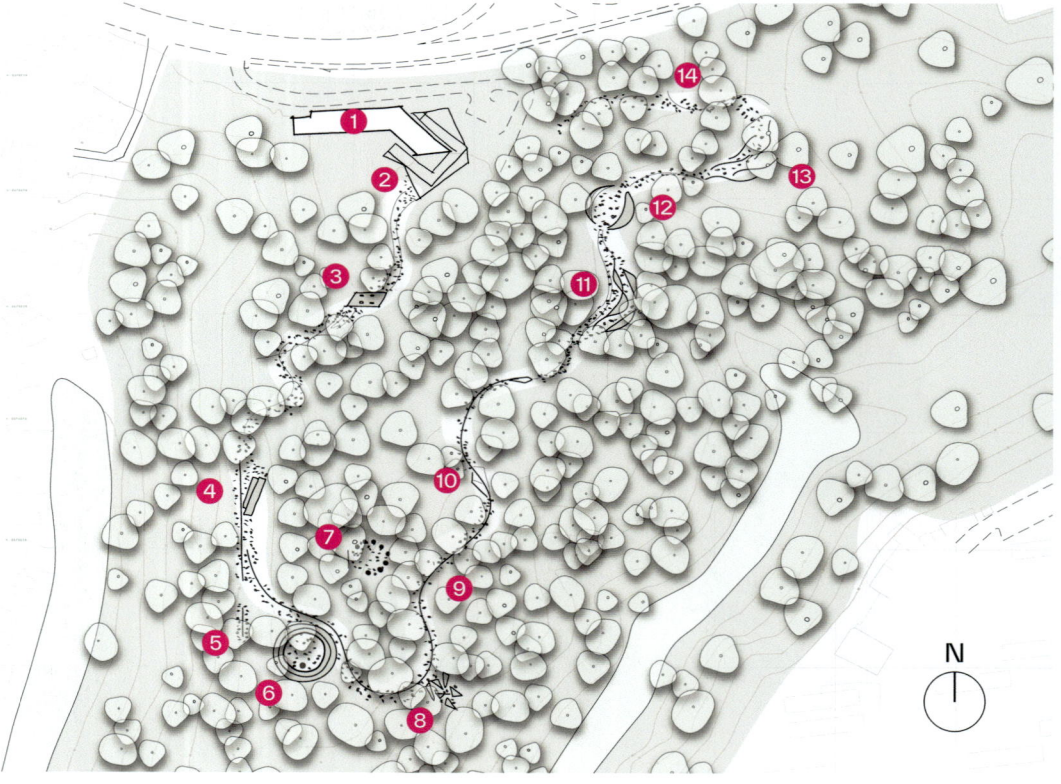

1 Exhibition space; 2 Starting point; 3 Exu – entrance; 4 Ogum – open paths;
5 Oxossi – inside the forest; 6 Oxalá – eyes to the sky; 7 Nanã; 8 Ossayn – master of the leaves;
9 Oxumaré – serpent; 10 Xangô – quarries; 11 Oxum – sweet waters; 12 Iemanjá – strong waters;
13 Iansã – winds; 14 Plant nursery.

(top) Heading towards the sky, *Oxala's* presence is associated to love and peacefulness. In his garden grounds a large circular shell-like structure surrounded by aromatic plants, inviting visitors to lie down and collectively watch the stars. (bottom) In Februrary 2019, the project was analyzed on site by Terreiro representatives, who suggested modifications such as the reallocation of certain gardens and botanical substitutions.

V. 'NATURE VILLAGE'
Middelfart (Denmark)

EFFEKT
with Anders Busse Nielsen and Björn Wiström
Design: yet to be realized

Client: Municipality of Middelfart

Naturbyen ('Nature Village') was envisioned to demonstrate how sustainable housing development can be combined with ambitious reforestation, increased biodiversity and circular resource thinking in suburban and peri-urban areas. The municipality-led 220-home residential expansion aims to contribute to Denmark's Natural Forest Programme goal of covering 20% (14% today) of its landmass with forest by 2100. A barren agricultural field in eastern Middelfart is developed as a cluster of small communities (15-20 homes) in a newly planted forest, uniting human and natural habitats. The agricultural land is restored through natural succession using well-established methodologies. The effect of a dense planting scheme with the nutrient rich agricultural soil accelerates the establishment of the woodland, which can take as little as 15 years – compared to 100-200 years with natural succession. The initial planting is very dense and creates an optimal microclimate that stimulates growth. After the first years of growth, the trees are progressively thinned out and new sub-layers of vegetation are planted to retain the optimal microclimate and increase biodiversity within the system. The new forest prioritizes edible crops such as fruits, nuts, root vegetables, mushrooms and so on, so that future residents can use the forest not only for recreational purposes but also as a source for nutrients. Similarly, local composting and small livestock will enable residents to nurture the forest. The forest protects groundwater, restores soil conditions, increases biodiversity and encourages social interaction. Housing clusters allow for differentiated housing types, community forms, forms of ownership and architecture, so as to create a diverse breeding ground for a broad composition of resident communities. Each home has direct access to small, private terraces – one oriented towards the courtyard and one towards the forest. The landscape and the spaces between the houses are shared between the residents. The forest areas are established on land that would otherwise typically be used as lawns and detached house gardens.

Phase 1 – Flower Meadow	Phase 2 – Tree Layer	Phase 3 – Bush Layer	Phase 4 – Herbaceous Layer

YEAR 0 ———— 1 ———————— 4 ———————— 7 ———————— 15 ————

Landscape succession stragegy

Agricultural land is reforested through residential development to create a healthy and socially connected neighbourhood. Natural succession is accelerated to establish a (sub)urban food forest. New housing typologies, with small footprints and wood construction, literally open up to the collective forest space.

VI. TRANSFORMING SOUTHBANK BOULEVARD
Melbourne (Australia)

TCL
Implementation: 2022   Client: City of Melbourne

Southbank is one of Melbourne's most densely populated suburbs and, prior to redevelopment, had the least amount of public open space per person in the municipality – 2.5 m² per person compared to the municipal target of 21m². Over 96 percent of Southbank's population live in apartments, limiting their access to outdoor space. These constraints are compounded by major arterial roads blocking many residents' access to existing open spaces. Transforming Southbank Boulevard was an ambitious, City of Melbourne-led project, which augments the vitality and livability of one of Melbourne's most densely populated suburbs. The project has transformed part of an arterial road into five new civic spaces, each delivering on the needs of the local community. The project has created an ecologically rich gateway, with over 400 new trees and climate-responsive, Australian native, understorey planting, at the heart of a burgeoning Melbourne Arts Precinct, Southbank Promenade, and Yarra River – Birrarung. Significantly, Transforming Southbank Boulevard has reallocated 22,000m² of public space for pedestrians, cyclists and children of all abilities, genders and ages. The City of Melbourne partnered with Yarra Trams to trial a 'green' tram track, along with trialing new tree species in response to the changing climate. The learnings from this trial have proved useful in further explorations of greening opportunities of tram tracks across the municipality. Integrating public art into the play space assisted in breaking out of the typical remit of 'playground', into a civic space that is not confined by a singular typology of 'artwork'/'plaza'/'playground'. Initially identified as an opportunity in the Southbank Structure Plan (2010), the transformation of Southbank Boulevard is the culmination of wide-ranging community and stakeholder consultation, strategic thinking, political will, complementary partnerships and ambitious design.

The project forcefully utilizes urban forestry to re-create a native-plant temperate biome in the city. Slow mobility and the interconnectedness of water and forest urbanism create a new public realm while mitigating consequences of global warming.

VII. MADRID METROPOLITAN FOREST (ZONE 4)
Madrid (Spain)

aldayjover architecture and landscape

Design: 2021-ongoing

Client: Madrid's Strategic Planning Office

The Metropolitan Forest is a circular park surrounding Madrid, a compact concentric metropolis of seven million inhabitants. This greenbelt is connected to larger regional parks in the mountains (to the north) and along rivers. In the context of global warming and increasing aridity in Spain, Madrid's Strategic Planning Office is investing in 4,300 hectares of forest to link the existing 27,700 hectares of large parks around Madrid. The 32,000-hectare Metropolitan Forest aims to improve the quality of life in the city and is part of the urban agenda for decarbonization. The Metropolitan Forest is divided into five zones led by different design teams that have together developed a palette of eight multi-strata forested landscapes based on soil conditions and water availability. Zone Four is a 1,250-hectare project at the intersection of the circular Metropolitan Forest and the north-south Manzanares River. In the absence of abundant rainfall (460 mm/ year), the city situated at the foothills of the northern snow-capped Guadarrama Mountains has historically been supplied by abundant groundwater qanats until the mid-nineteenth century. The project acknowledges this tradition and recognizes the importance of water infrastructure. It magnifies the water cycle in three new and recovered historical aspects: 1) increased floodable areas (deconstructing dikes), 2) a network of natural and artificial green-blue infrastructures, and 3) an increased volume of used water and sludge from sewer treatment plants to support a palette of forested landscapes. The project contributes to slow mobility, civic nodes and connectivity infrastructures between the city and the Metropolitan Forest.

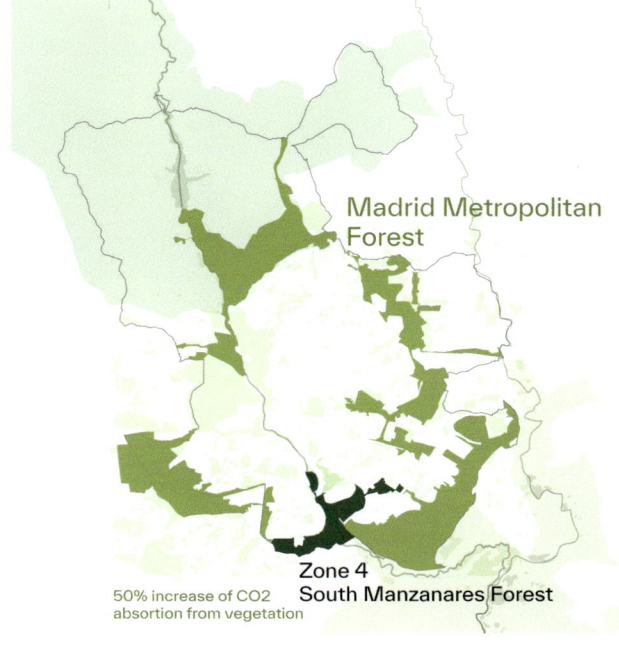

50% increase of CO2 absorption from vegetation

Zone 4
South Manzanares Forest

The Metropolitan Forest surrounding Madrid highlights two ambitions: the continuity of existing parks and the connectivity of this large greenbelt outwards to existing larger regional parks. Situated at the intersection of the 75-km-long circular Metropolitan Forest and the north-south Manzanares River crossing the city is the 1250-hectare Zone 4 (in darker green).

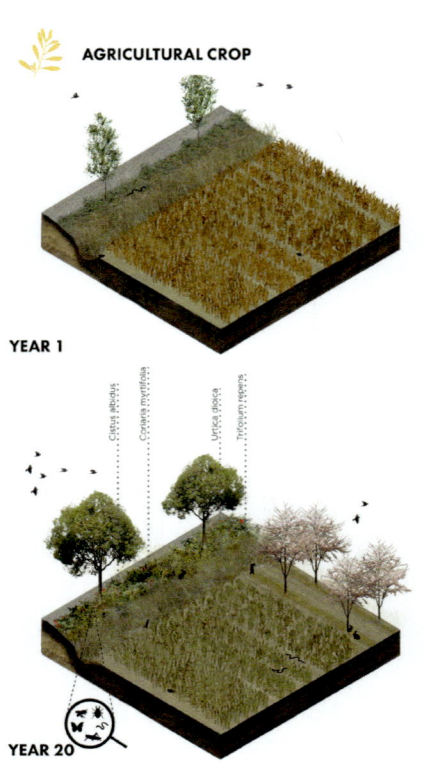
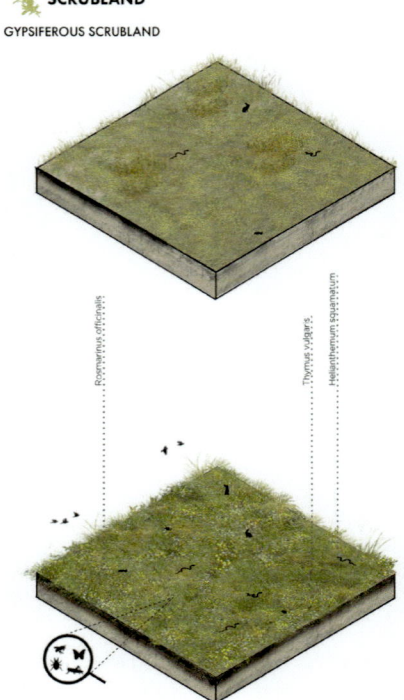
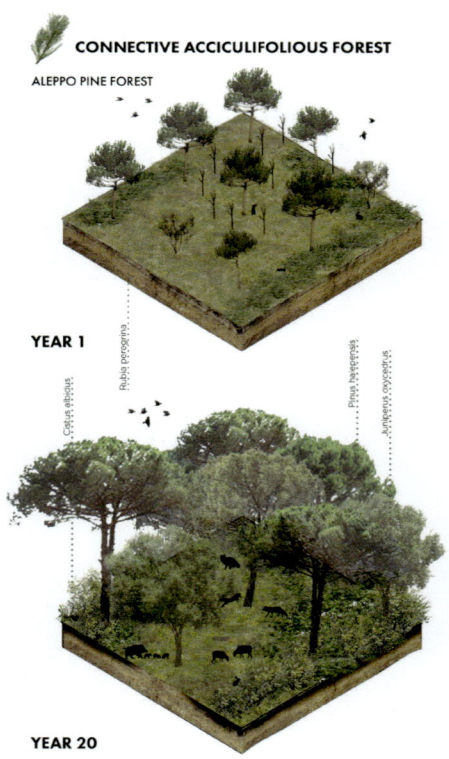

DRY STREAM
HALOPHYTE TAMARISK

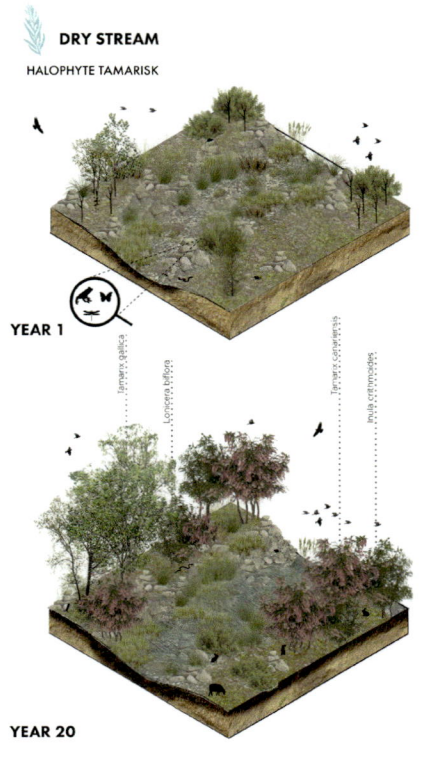

YEAR 1

YEAR 20

URBAN TRANSITION PARK
LOW DENSITY POPLAR FOREST AND SCLEROPHYLLOUS SHRUBS

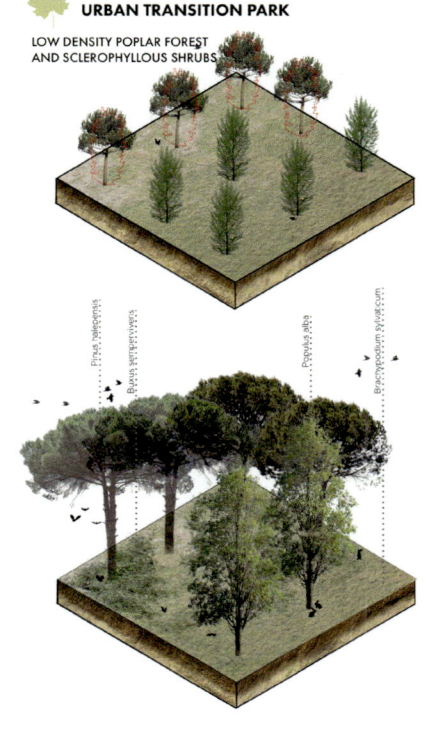

WETLAND
EUTROPHIC STAGNANT FRESH WATER

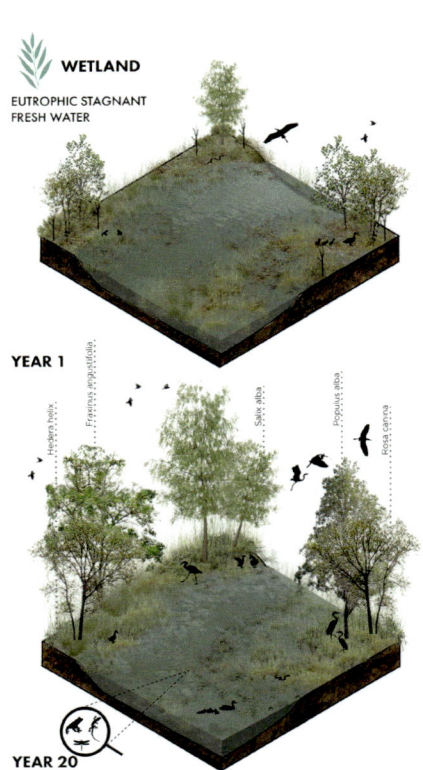

YEAR 1

YEAR 20

RIPARIAN FOREST
WILLOW ADN POPLAR GALLERY FOREST

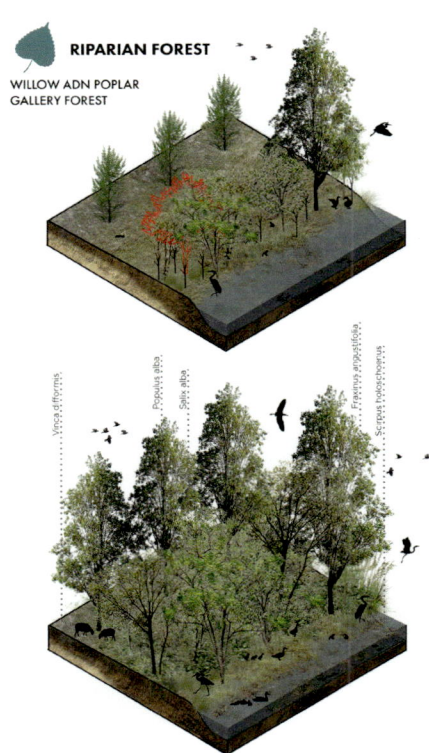

Planta General

Along the north-south Manzanares River, four strategies are used: 1) an increased floodability regime along with larger areas of riparian forests, 2) increased reuse of regenerated waters from sewer treatment plants for drip irrigation, 3) a monumental wetland, and 4) the recovery of orchards along the river. The western wing runs along the M45 highway. The rain-based landscape will transform into a matrix of pine forest ecosystem and agricultural fields. The continuity of the Bulera arroyo will allow the ecological diversity of flora and fauna to grow over time. Along the dry La Gavia arroyo in the east, the project proposes low-cost management of the threatened Mediterranean gypsum scrub habitat, which is of high ecological interest to the EU28. Grazing, fire, excessive wind and timber harvesting have been the main drivers of this habitat's degradation in the past, replaced today by severe drought, excessive slopes, urbanization and extreme soil conditions. Changing unfavourable conditions through the insertion of drip infrastructure supplied by regenerated waters will propel these vulnerable habitats towards their natural succession.

Historically, compact Mediterranean cities like Madrid have had an aversion to trees. The car revolution of the 1960s-70s eliminated several eighteenth- and nineteenth-century forested boulevards. Fortunately, recent initiatives led by the city are changing course, allowing both urban forestry and forest urbanism to flourish.

FOREST URBANISMS EXPLORATIONS

I. NATIVE, EXOTIC OR FUTURE-PROOF? NAVIGATING URBAN TREE PLANTING IN THE BRUSSELS MAELBEEK VALLEY

Björn Bracke, Koenraad Danneels, Marlène Boura

In January 2022, the Brussels region and the municipality of Ixelles announced a project to green the Place Flagey, one of the largest public squares in the city. The square, whose previous reconstruction was only completed in 2009, is deemed too mineral and insufficiently adapted to the current challenges concerning water management, the urban heat island, and the decline of urban biodiversity (BX1 2023). Moreover, the Place Flagey is located in the Maelbeek Valley, part of the urbanized nineteenth-century belt of Brussels which the regional government has selected as an important blue-green corridor in the sustainable development plan for the Brussels Capital Region. The winning design proposal of the 2022 call, won by Kollektif landscape and Fallow architects, envisages 60 new trees and the removal of approximately 2000 m² of paving in favour of new patches of vegetation. This urban planting project is an interesting case to reflect on the choice of tree species and, more specifically, the use of native or non-native species. The original design of Latz and Partners in 2009 included a group of more 'ornamental' exotic trees in the north end of the square and a group of native trees in the south end, on the side of the ponds of Ixelles. This group, including willows (*Salix alba*), were a nod to the typical wetland vegetation in the Maelbeek Valley. Today, in line with the policy of the governmental agency Brussels Environment, the new design envisages 90% native trees. Championing the rediscovery of a premodern valley, the ecological restoration agenda echoes planning policy documents that recognize the Maelbeek Valley as an urban ecological corridor where the city has been promoting the use of indigenous tree species (Fontaine & Gryseels 2016; Brussels Environment 2017). Yet this specific historical and ecological framing seems to ignore the presence of nineteenth-century historic gardens and parks that, contrary to the premodern landscape, bolster a wide range of remarkable, often non-native, species, many of which enjoy protected heritage status. In this paper we want to explore how this nativist agenda relates to the existing urban tree stock, specifically in the context of changing climatic conditions.

Questions linked to the 'greening' of Place Flagey illustrate the ambiguity and complexity that the current ecological urban policies, as well as related nativist approaches, are creating in relation to nineteenth- and twentieth-century notions of urban design, grounded in architectural and horticultural traditions. Many authors have already challenged the problematic terminology of

Towards a greener Place Flagey. The complex growing conditions atop of an underground parking structure, the historical identity of the built environment and its biophysical location in the Maelbeek Valley offer a multitude of perspectives on the 'greening operation'.

'native', 'exotic' or 'invasive', terms that are heavily value-laden and based on implicit cultural values and assumptions (Kendle & Rose 2000). Bruno Latour has more generally pointed towards fundamental questions concerning ecological principals that aim to be holistic but inevitably become isolating and reductionist (Latour 1998; Kendle & Rose 2000). While the categorization of native/non-native is still championed in ecological planning methods, in his recent book *Natura Urbana*, Matthew Gandy has argued that there are three potential problems with these kind of policies, which we can link to the Maelbeek case: (1) there is "the relatively arbitrary choice of ecological baselines", as exemplified by the choice for pre-modern landscape types in the Maelbeek in the Place Flagey design; (2) the "presence of harmless novel ecological assemblages" is criticized, while nineteenth-century parks that house exotic plant species are actually of important value; and (3) the "potential advantages of new socio-ecological interactions under changing environmental conditions" are ignored (Gandy 2022: 218). Gandy also argues that climate change is "unsettling any meaningful distinction between 'native' and 'nonnative' species" as temperature changes impact the survival chances of certain tree species (Gandy 2022: 218).

In this contribution, we want to illustrate this (often abstract) debate in urban ecology discourse with a concrete case in Brussels by analyzing the existing tree demography in the Maelbeek Valley and confronting these datasets with ecological planning policies and discourses. We will use this data research to question pastoral notions of the 'lost' valley landscape in Brussels. We also want to take stock of the ways in which current policy-making is disregarding existing plant and animal assemblages by a reductive focus on native, or red-listed species. Current multi-species approaches in literature on planning and design call for a more inclusive urbanism practice that takes existing multi-species assemblages into account (Houston et al. 2018; Bracke et al. 2022). Current ecological visions for tree planting in the city run counter to such a caretaker position for existing plants while, somewhat surprisingly, that it is only through heritage policies that exotic tree species are protected. To understand the approach of Brussels planning practices, we will first analyze the Brussels policy context. We will then present an in-depth analysis of the existing tree demography through the construction of a tree dataset. Finally, we will discuss the outcomes and link them to the debate around nativist discourse in landscape architecture and urban planning. By mapping the tree canopy *as is* – and therefore raising attention to the actual trees present on site – we hope to upset and subsequently redirect ecological planning policies towards a planning approach that works with, rather than against, tree populations already present in the valley.

The Maelbeek Valley as a green corridor

Historically, the Maelbeek River is a tributary of the Senne River that flows through the Brussels city centre. The stream was vaulted over during the nineteenth century (Leloutre 2018). Today, the whole area is heavily urbanized and characterized by nineteenth-century residential neighbourhoods, parks and the European Union headquarters, including large institutional office buildings. The valley, while completely urbanized, plays an important role in the project of the Green Network of Brussels. As many in Western cities, Brussels devised an ecological vision for its green spaces. With the "green network map" of the regional plan for sustainable development (2013) and the *Plan Nature* (2016), the region has been developing strategic documents and action points to support the development of a coherent ecological structure for the region. Both the regional plan for sustainable development and the *Plan Nature* describes the Maelbeek Valley as a green corridor between the Zonian Forest in the southern parts of the city and the Parc Josaphat in the north (Fontaine & Gryseels 2016). The *Plan Nature* echoes current international policy discourses in which nature development should not only focus on natural reserves, but also stress the important role that cities can play in safeguarding biodiversity (European Commission 2020). One of the goals of the *Plan Nature* is to "provide for a significant planting of trees in the coming years" and to "prioritize the planting of native species in these projects" (Fontaine & Gryseels 2016: 59). While the *Plan Nature* declares only to combat invasive exotic species and clearly states that most exotic species are not a problem for biodiversity, the list of recommended

Projection of the Maelbeek Valley on the Green Network of the Regional Sustainable Development Plan. On the ground, this green-blue axis is barely visible; the Maelbeek was vaulted over in the nineteenth century, and the valley is highly urbanized.

plants from Brussels Environment includes only native tree species (Bruxelles Environnement 2017). These measures are symptomatic of a wider wave of 'nativist' approaches introduced into urban planning in recent decades. Many sustainability indicators and assessment tools (e.g., BREEAM) also systematically give higher scores to native planting strategies, regardless of context.

Mapping the trees of the Maelbeek Valley

This recently adopted nativist approach – influenced by principles from conservation biology and propelled by the urban green services of 'Brussels Environment' – sharply contrasts with the rationales of other urban policies. For instance, in the transportation departments a managerial logic of urban tree care predominates while heritage administrations in recent decades have assigned protected status to many exotic trees (Leclercq & Campanella 2013), promising to replant those same species if these "remarkable trees" die. In order to understand how these dilemmas apply to the existing situation, we think it is important to understand what tree population is actually present in the Maelbeek, so we mapped and analyzed its existing tree population. This exercise was done for an area of 1.38 km², along a 3.75 km axis between Ambiorix Square and La Cambre Abbey. We created a database of the trees of Brussels using the two inventories that were carried out in the Brussels region and completed the characteristics on the tree species thanks to additional sources. Brussels Mobility surveyed most of the street trees, while Urban Brussels started to survey remarkable trees in 2002 (Urban Brussels 2022; Brussels Mobilité 2022). Neither of these inventories is exhaustive, but combined they provide a geodatabase of 44,813 trees at the regional level. It considers trees along streets, boulevards and avenues for the former, and in green spaces – public or private – for the latter. The figure displays the spatial distribution of the trees inventoried with their type (indigenous, exotic or hybrid).

Indigenous or native trees	Trees occurring within their natural range (past or present) and dispersal potential (i.e., within the range they occupy naturally or could occupy without direct or indirect introduction or care by humans)
Hybrid trees	Trees that are the result of crossbreeding or hybridization between different species or varieties of trees
Exotic, nonnative or alien trees	Trees occurring outside of their natural range (past or present) and dispersal potential (i.e., outside the range they occupy naturally or could not occupy without direct or indirect introduction or care by humans)
Invasive trees	Exotic tree species which become established in natural or semi-natural ecosystems or habitat, are an agent of change, and threaten native biological diversity

Definition of terms (based on International Union for Conservation of Nature 2000)

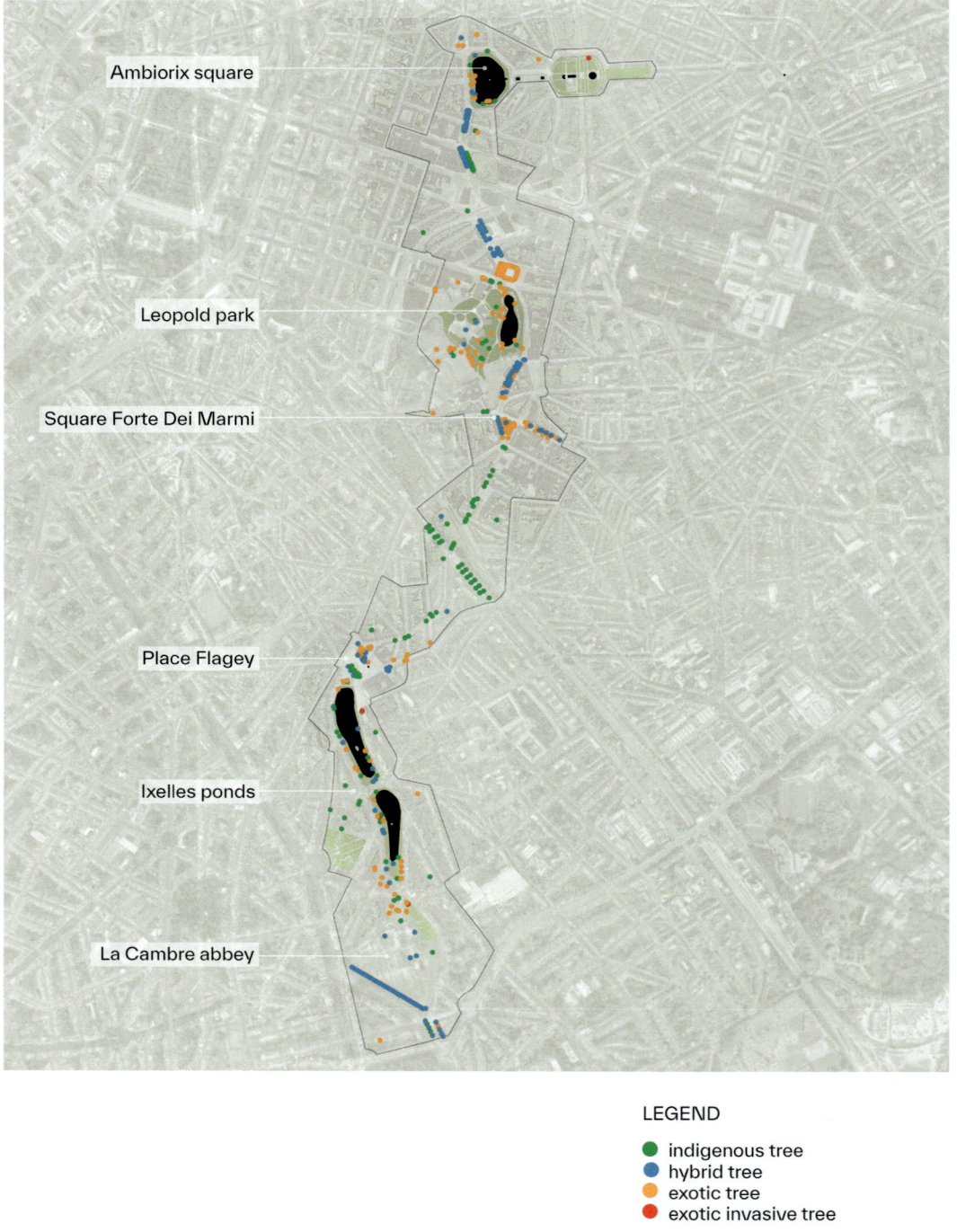

Map of the trees in the Maelbeek Valley with the type of species. Only 23% of trees are indigenous, and particularly the historical parks have a strong exotic character.

Species indigenous to Brussels are listed by Brussels Environment and the City of Brussels (Villes de Bruxelles 2017; Bruxelles Environnement 2017). In case of conflicting information, we used the Belgian species list (Royal Belgian Institute of Natural Sciences 2021). From the project LIFE AlterIAS, a list of invasive species for Belgium has been established (AlterIAS 2013). For the invasive species we followed the Invasive Species Environmental Impact Assessment (ISEIA) protocol (Biodiversity. be 2009), whose list is regularly updated and made available via the Harmonia Database (Belgian Forum on Invasive Species 2021). Hybrid species are considered separately as they are never regarded as indigenous, nor do they appear to be assessed by the ISEIA protocol or European Red list. A species is considered exotic if it is not listed as indigenous or hybrid. Endangered species were identified from the European Red List (Rivers et al. 2019) provided by the International Union for Conservation of Nature and Natural Resources (IUCN). It is important to note that some species exotic to Belgium or the Brussels Region may also be considered as endangered within Europe.

In total, 550 trees were inventoried in the Maelbeek Valley. Fifty-seven species were identified, of which ten species represent 79.27% of the tree population. The species *Platanus x acerifolia*, commonly called London plane, represents more than one fifth of the tree population. Of the ten most widespread species, four are exotic, but none are invasive, and three are hybrids. *Aesculus hippocastanum*, commonly known as the horse chestnut, is both exotic in Belgium and qualified as vulnerable in Europe.

Type	Family	Species	Cat.	Observations (%)
Indigenous (16.36)	*Fagaceae*	*Fagus sylvatica*	Least Concern	10 (1.81)
	Sapindaceae	*Acer platanoides*[1]	Least Concern	47 (8.45)
	Sapindaceae	*Acer pseudoplatanus*[2]	Least Concern	33 (6.00)

Type	Family	Species	Cat.	Observations (%)
Exotic (27.81)	Altingiaceae	Liquidambar styraciflua	None	32 (5.81)
	Hamamelidaceae	Parrotia persica	None	45 (8.18)
	Sapindaceae	Aesculus hippocastanum[1]	Vulnerable	52 (9.45)
	Ulmaceae	Zelkova serrata	None	24 (4.36)
Hybrid (35.09)	Malvaceae	Tilia x euchlora	None	70 (12.72)
	Platanaceae	Platanus x acerifolia	None	111 (20.18)
	Sapindaceae	Aesculus x carnea	None	12 (2.18)

Top 10 tree species in the Maelbeek Valley. Exotic and hybrid trees are predominant ([1] naturalization, [2] naturalized).

When looking at the entire tree population, we found that only 16 of the 57 tree species are indigenous, while exotic species are predominant, including two invasive species. The five hybrid species amount to 35.81% of the tree population, and the exotic species represent 41.44% of all trees inventoried. Two exotic invasive tree species are found in the valley: *Ailanthus altissimsa* (four individuals from which two have been cut down) and *Robinia pseudoacacia* (one individual). *Ailanthus altissimsa* (category A2) is blacklisted in Belgium: it is considered to have a high environmental impact, but it is spread in a restricted range across Belgium. *Robinia pseudoacacia* (category B3) belongs to the watch list in Belgium: it is considered to have a moderate environmental impact and to be widespread throughout Belgium.

Type	Number of species	(%)	Number of trees	(%)
Indigenous	16	(28.07)	125	(22.72)
Exotic	34	(59.64)	223	(40.54)
Exotic invasive	2	(3.50)	5	(0.90)
Hybrid	5	(8.77)	197	(35.81)
Total	57		550	

Overview of the different groups of trees mapped in the Maelbeek Valley. Only 22,72% of the existing trees were found indigenous.

The Maelbeek Valley: Exotic, biodiverse or future-proof?

In total, almost 80% of the existing species in the Maelbeek Valley (of which 35% are hybrids) are not indigenous, contrary to what policies hope to develop in this ecological corridor. Indeed, as Kendle and Rose (2000) argued, non-natives have played an important role in the development of the environment, and it is neither possible nor desirable to ignore or erase this history. Many of the exotic trees in the Maelbeek Valley are situated in important historical parks and squares such as the Leopold Park, Forte Dei Marmi Square or Jean Rey Square. The Leopold Park, for example, was conceived as a zoological and botanical garden to bring the exotic world to Brussels inhabitants in the nineteenth century and was later used by universities to extend their campus (Lambrechts 2015). This park, along with many throughout the valley, is filled with "remarkable trees", as defined by governmental regulations (Bruxelles Environnement 2011). The maintenance of these parks is fraught with dilemmas, as heritage, mobility and ecological standpoints often run counter to one another (Leclercq and Campanella 2013). Today the regions' heritage policy aims to replace the remarkable, often exotic, trees with the same species in case of death. Also, it is peculiar that there is an increasingly strong condemnation of non-native species in nature policy documents just at the time when environmental and climate conditions are changing (IPCC 2019). Wallonie Environnement SPW, for example, identified indigenous tree species and genera which are or will face challenges due to the ongoing climate change and forecasts over the coming decades, as provided by the Intergovernmental Panel on Climate Change (IPCC). The Royal Forest Society of Belgium is currently testing the resilience of indigenous species with different provenances as well as exotic species with a higher potential to handle the climate change context in Belgium (Royal Forest Society of Belgium 2022). A selection of tree species is now being assessed based on a set of criteria: "their adjustment to the climate now and in the future, their ability to resist pests (insects) and pathogens (diseases, fungi), the productivity and quality of the wood for timber production, their effect on biodiversity (ability to host flora and fauna and the risk of invasion)" (Royal Forest Society of Belgium 2022). These elements support a better understanding of the possible demographic evolution of trees in Belgium. Among the indigenous species they considered, some are also present in the Maelbeek Valley. *Tilia cordata* (small-leaved linden, two individuals), could be favoured by the increase of temperatures, while F*agus* species, of which eight are present in the valley, are amongst the most at risk due to their important level of sensitivity to heat waves, drought and soil waterlogging, in addition to their sensitivity to hydric deficit and stability against winds (Quévy 2017; Royal Forest Society of Belgium 2022).Similarly, but to a lesser extent, F*raxinus* species and A*cer pseudoplatanus* are sensitive to the current and coming climatic changes. Of the 33 A*cer pseudoplatanus* (Plane tree) inventoried, 27 remain in the valley, and they are sensitive to heatwaves, drought and hydric deficit (Royal Forest Society of Belgium 2022; Quévy 2017). Out of the 28 alternative "future-oriented" species listed by the Royal Forest Society of Belgium, eight are present in the Maelbeek Valley. Two of them are indigenous, and six are exotic. In addition, at least three of the indigenous species are at risk.

    These observations illustrate how labels like native, exotic, and even invasive are fluid and how reductionist approaches towards urban ecology do not recognize and face the ultimate uncertainty of future climate changes. The aspirations for nature conservation and promotion of native species, as promoted by the *Plan Nature*, are based on the ideas of stability and continuity while recent studies show that ecological systems are not stable (Rotherham 2017). Existing and emerging urban ecologies are always hybrid, and might cause disruptions of native ecosystems and disappearance of local species (Rotherham 2017; Branquart 2012). The mixing of species is occurring at a rate unprecedented in history and is most easily observed and recognized in the urban environment (Rotherham 2017). Since climatic conditions are changing across the globe (IPCC 2019), these shifts already having significant impacts on the status of trees according to their positive or negative responses to water and heat stresses (Quévy 2017). This also means tree populations risk disappearing in their native regions, and the migration of species, whether natural or assisted, is an important topic in forestry research (Royal Forest Society of Belgium 2022; Quévy 2017).

A laboratory for hybrid ecosystems

From our analysis of the tree stock, we can conclude that the vast majority of the existing trees do not fit in with the current policy objectives for biodiversity. The case of the Maelbeek Valley reveals three significant issues. Firstly, the cultural and historical background of the urban area has resulted in a predominantly exotic and hybrid population of trees. Secondly, changed climatic conditions such as increasing temperatures have a harmful impact on a large number of the indigenous tree populations. Of the 125 indigenous species inventoried in the Maelbeek Valley, 47 are considered at risk due to climate change. Lastly, various exotic species, some of which are already present in the valley, can become part of mitigation strategies to better counter urban heat island effects and biodiversity loss in the city (Quévy 2017; Royal Forest Society of Belgium 2022).

Beyond the discussion on the native or exotic status of trees, the analysis particularly shows the problematic application of reductionist methodologies and labels within the domain of urban ecology. As a reaction against these technocratic or managerial approaches, Bruno Latour advocated for a more relational paradigm and "profound reassessment of our interactions with the natural environment" (Latour 1998). In the same vein, multi-species approaches have rapidly gained traction in recent years among scholars, policy-makers and designers. This method's appeal brings us back to the Flagey greening project. Like many contemporary landscape designers, the design team portrays a more relational conception of urban nature, illustrating the interactions of the trees with the environment and urban fauna. However, these representations often focus on new 'designed assemblages' and are not grounded in a profound understanding of how existing multispecies relations evolve. To do this effectively, urban vegetation would have to be documented and monitored in more vigorous and site-specific ways.

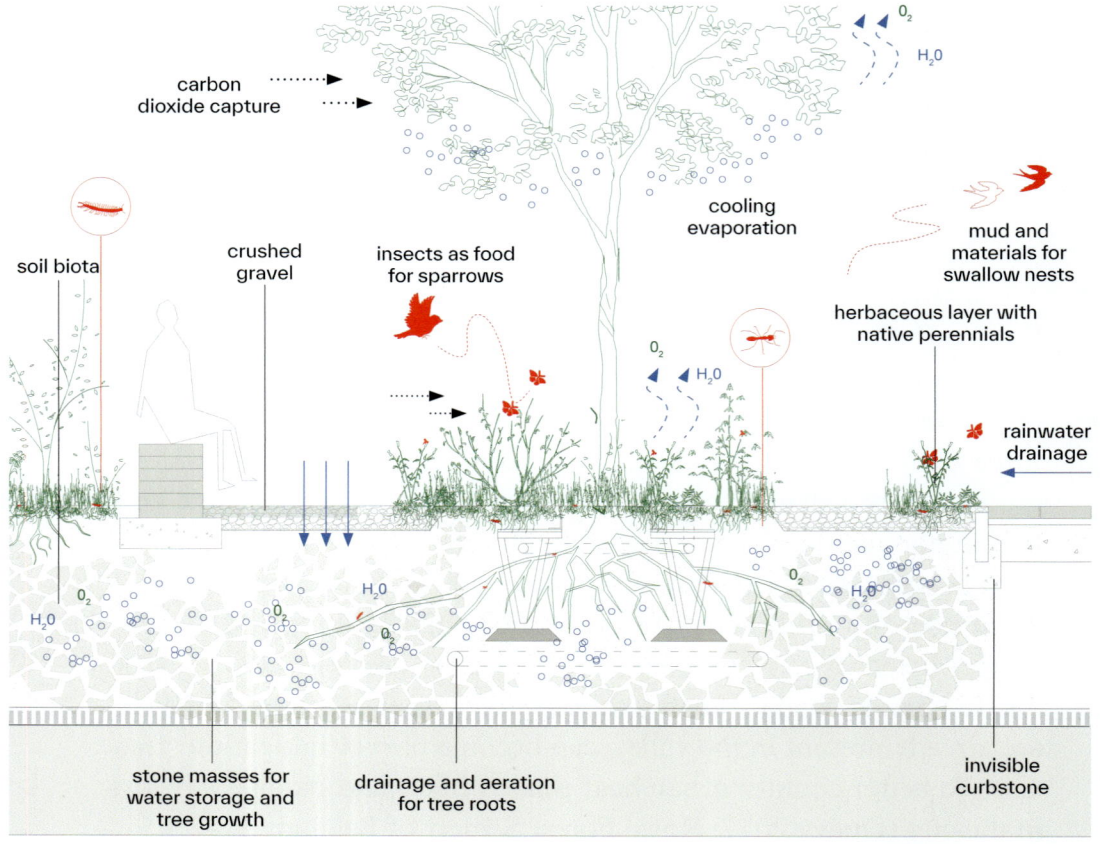

Section explaining the multiple benefits and relations of the new plantings for Place Flagey. More relational and multi-species representations have been popularized in recent years in landscape and urban design.

This brings us to the role of data and research in urban ecology. Who is managing and developing the data, and what policy agenda does it support? The data sets used for this paper were initially not developed to foster reflections on urban ecology. First, the tree databases mobilized in this paper are generated and managed by different departments, Brussels Mobility and Urban Brussels, two governmental agencies that have a transportation and heritage agenda. Second, the research we used to reflect on the suitability of trees in changing climatic conditions is based on studies initiated by arboriculturists, with clear economic aspirations. If we want to implement a more relational or multi-species approach, we need other data to support this (soil biota, root systems, growth, flowering behaviour, reproduction, urban fauna interactions etc.). Maybe there is a role to be played by citizen science or crowdsourced initiatives in collecting more observational and relational data on tree species and their habitat?

Our contribution started with a reflection on dominant ecological discourses in Brussels planning policy and linked it with ongoing debate in ecology and forestry research. We argued to critically examine the value judgements associated with labels like native or exotic. Ecological design often draws on discourses of 'ecological restoration' and other types of historically framed cultural landscape to imply a sense of continuity with the past (Gandy 2015). Our analysis showed that we will have to cope with the fact that future urban ecologies are increasingly hybrid fusions of species. Rather than reconstructing its ecological past, we believe that cities should be conceptualized and managed as laboratories for new hybrid ecosystems where intimate mixes of now native and exotic species are tested and co-evolve towards new ecologies. Therefore, we need a good understanding and anchoring of these environmental policies in the historical context of the city since natural heritage is part of an ecological and urban identity. As a dynamic and cosmopolitan city, Brussels can be seen as a laboratory for the emergence of these 'new ecologies' or hybrid ecosystems where novel interactions and dependencies are formed. The "wonderful cacophony and experiment of intermingling" of plant species, as Gilles Clement describes it, responds to the multi-ethnicity and openness that has been one of the driving forces of thriving cities throughout history (Clément 2015; Konijnendijk 2018).

Disclosure statement

The first author was involved in the design of this proposal and the further elaboration of the greening project for the Flagey Square as partner of Kollektif landscape. The chapter was written in the context of the CO-HABITAT research project, funded by Innoviris.

References

AlterIAS. 2013. *List of Invasive Plants in Belgium*. Gembloux: Laboratoire d'Ecologie, Université de Liège.

Belgian Forum on Invasive Species. (2021). Harmonia Database. *Belgian Biodiversity Platform*.

Biodiversity.be. (2009). *Guidelines for Environmental Impact Assessment and List Classification of Non-Native Organisms in Belgium*. Version 2.6.

Bracke, B., Bonin, S., Notteboom, B., Leinfelder, H. (2022). A Multispecies Design Approach in the Eure Valley. Three Lessons from a Design Studio in Landscape Architecture. *Cahiers de La Recherche Architecturale, Urbaine et Paysagère*no, 14 (April 12).

Branquart, E. (2012). Arbres et Arbustes Exotiques: Une Nouvelle Vague d'envahissuers? *Forêt Wallone*, 120, 42–48. Gembloux : Cellule Espèces Invasives.

Bruxelles Environnement. (2011). *Le Parc Léopold. Un Musée de l'architecture En Plein Air*. Brussels: Bruxelles Environnement.

Bruxelles Environnement. (2017). *Espèces Végétales Indigènes et Conseillées. Info Fiches Espaces Verts – Biodiversité*. Brussels: Bruxelles Environnement.

Bruxelles Mobilité. (2022). Inventaire Georéférencé Des Arbres. *Mobility.Irisnet.Be*.

BX1. (2023). Audrey Lhoest: "Il Est Grand Temps de Verduriser et de Planter Des Arbres à Ixelles," June 1.

Clément, G. (2015). *The Planetary Garden and Other Writings*. Philadelphia: University of Pennsylvania Press.

European Commission. (2020). *Natura 2000 in Cities*. Luxembourg: Publications Office of the European Union.

Fontaine, F., & Gryseels, M. (2016). *Plan Nature, Plan Régional Nature 2016-2020 En Région de Bruxelles-Capitale*. Brussels: Les fiches d'information sur les espaces verts de la Région de Bruxelles-Capitale.

Gandy, M. (2015). From Urban Ecology to Ecological Urbanism: An Ambiguous Trajectory. *Area* 47, 2 (June 23), 150–154.

Gandy, M. (2022). *Natura Urbana : Ecological Constellations in Urban Space*. Cambridge, MA: MIT Press.

Houston, D., Hillier, J., MacCallum, D., Steele, W., Byrne, J. (2018). Make Kin, Not Cities! Multispecies Entanglements and "Becoming-World" in Planning Theory. *Planning Theory*, 17, 2 (May 1), 190–212.

International Union for Conservation of Nature. 2000. *IUCN Guidelines for the Prevention of Biodiversity Loss Caused by Alien Invasive Species*. Gland [Switzerland]: IUCN.

IPCC. (2019). *Global Warming of 1.5°C. An IPCC Special Report on the Impacts of Global Warming of 1.5°C above Pre-Industrial Levels and Related Global Greenhouse Gas Emission Pathways, in the Context of Strengthening the Global Response to the Threat of Climate Change*.

Kendle, A.D., & Rose, J.E. (2000). The Aliens Have Landed! What Are the Justifications for "Native Only" Policies in Landscape Plantings? *Landscape and Urban Planning*, 47, 1–2 (February), 19–31.

Konijnendijk, C. (2018). *The Forest and the City: The Cultural Landscape of Urban Woodland*. Cham: Springer.

Lambrechts, W. (2015). De Brusselse Dierentuin. Sociale Ontmoetingsplaats Voor de Burgerij in Een Exotisch Kader. *Erfgoed Brussel*, 17, 82–93.

Latour, B. (1998). To Modernize or to Ecologize, That Is the Question. In Castree, N., & Willems-Braun, B. (eds.), *Remaking Reality: Nature at the Millenium* (pp. 221–242). London and New York: Routledge.

Leclercq, C., & B. Campanella. (2013). De Inventaris van de Merkwaardige Bomen. Brussels: Erfgoed Brussel.

Leloutre, G. (2018). Brussels ' Heterogeneity and Fragmentation via Topography. In Ranzato, M. (ed.), *Water vs. Urban Scape* (pp. 285–306). Berlin: Jovis.

Quévy, B. (2017). Le Changement Climatique et Ses Impacts Sur Les Forêts Wallonnes. Recommandations Aux Décideurs et Aux Propriétaires et Gestionnaires. Jambes: Wallonie Environnement SPW.

Rivers, M., Beech, E., Bazos, I., Bogunić, F., Buira, A., Caković, D., Carapeto, A., et al. (2019). *European Red List of Trees. European Red List of Trees*. Cambridge, UK and Brussels, Belgium: IUCN, International Union for Conservation of Nature.

Rotherham, I.D. (2017). *Recombinant Ecology. A Hybrid Future?* Cham: Springer.

Royal Belgian Institute of Natural Sciences. (2021). *Belgian Species List. OD Nature, Royal Belgian Institute of Natural Sciences*. Brussels: Institute of Natural Sciences.

Royal Forest Society of Belgium. (2022). *Trees for Future*. Gembloux: Société Royale Forestière de Belgique.

Urban Brussels. (2022). *Inventaire Du Patrimoine Naturel (Arbres Remarquables et Sites)*. Brussels : Urban Brussels.

Villes de Bruxelles. (2017). *Liste Des Plantes Indigènes Utiles Pour La Biodiversité*. Brussels: Eco-Conseil.

II. FORESTING A *CHAR* IN THE BRAHMAPUTRA VALLEY IN ASSAM, INDIA

Swagata Das, Kelly Shannon, Bruno De Meulder

Since the independence of India, the relatively unexplored territories of Assam and the Brahmaputra Valley were considered untouched lands and Indigenous territories to be opened for development. The colonial forest policy in India encouraged settlements in peripheral areas, through systems like *begar* or unpaid labour (Sharma & Sarma 2014). Such systems granted settlers restricted rights to specific forest land without legal tenure. The colonial policies that ordered nature also functioned as control policies of minorities (Scott 2020). In the 1970s, the post-colonial Indian government continued to assert control over forests by implementing new conservation policies which sought to transform them into people-free zones (Cremin 2011). The interplay between the Indigenous system and the formal legal framework inherited from the British colonial authorities reflects distinctive values rooted in diverse cultural contexts and worldviews. While the formal system revolves around private and individual property, the Indigenous system was founded on the management of common property resources (Gadgil & Guha 1995).

In recent years, there has been a notable shift in attitudes towards forest-related practices and local knowledge, acknowledging the potential of forest communities to inspire innovative management approaches. This research delves into the transformation of Indigenous territories, exemplified by a plantation initiative on a *char*, or river-based sandbar. Initially conceived to assert authority over dynamic riverine territories, the initiative evolved into a community-driven revival of forests orchestrated by the Mising ethnic group.

To comprehend the dynamics and role of local communities in sustaining forests, it is crucial to "avoid standardization of 'local knowledge' into a few simple 'blueprint' techniques of forest production and the idealization of indigenous people as the universal, legitimate and talented stewards of the forest for the future" (Michon 2005: 175). Forests have long been studied as prime examples of environmental activism against ecological imbalance (Bhattacharya 2018; Jain 1984; Barbosa 2015). This study refrains from celebrating a single individual/community's efforts to resist imposed systems and combat global warming. Instead, it highlights an evolving world of relations between nature and society, presenting an alternative approach to spatial designs and transformation based on the role of nature (forest) in the Mising worldview. Rather than portraying the case as an idealistic model to replicate, the study presents it as a contemporary and unconventional settlement approach grounded in nuanced response to ever-evolving realities. As such, it stands as a relevant and potential stepping stone to forest urbanism.

Shifting territories

With one of the highest sediment yields in the world – 852.4 tons/km^2/year (Lahiri & Sinha 2012) – the Brahmaputra is a transnational river originating in Tibet. As it enters India, the sudden flattening of its slope and joining of tributaries carrying massive quantities of sediment result in an oscillating braided pattern. The braided river forms a network of channels interspersed with islands known as *chars*. The braided nature of the river, whose morphology undergoes frequent changes due to natural causes like earthquakes, annual flooding and man-made interventions, such as embankments and dams which tame the river. *Char,* as a landform, is born from this dynamism.

Throughout history, one finds reference to *char-chaporis* (river islands and riverbanks) and the need to optimize the economic and political benefits of the untamed territories that are "liable to annual inundations or irruptions of new channels and are abandoned to a wild overgrowth of reeds or grass-jungles, incapable of permanent habitation or cultivation" (Hunter 1885: 346). However, the shifting landscapes are inhabited and cultivated,

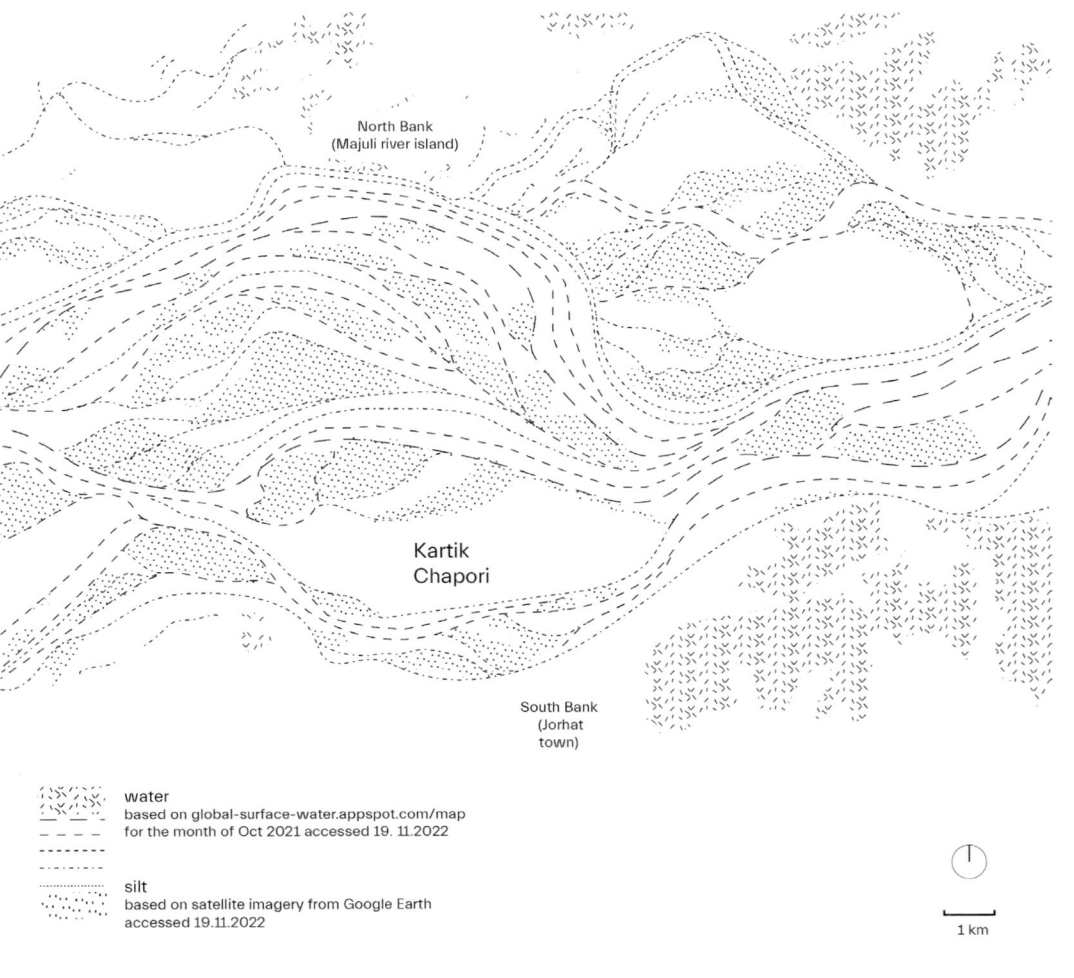

Gradients of the fluid landscape. The transition from lines (water) to dots (silt) illustrates an ever-evolving domain with continuously shifting degrees of wetness/gradients of so(i)lidness.

especially by the Mising community, often referred to as "river people" (Thakur 2021). The Misings are of Tibeto-Burmese origin and have developed a mobile lifestyle seeking fertile grounds, adjusting to floods, altering channels and inhabiting the *char-chaporis*. The studied *char* – Kartik *chapori* – formed through years of shifting water and silt, along with a Mising village comprising of *Okum* (Mising stilt houses), built four to five feet above the ground. It is within this shifting landscape that an expanding forest exists.

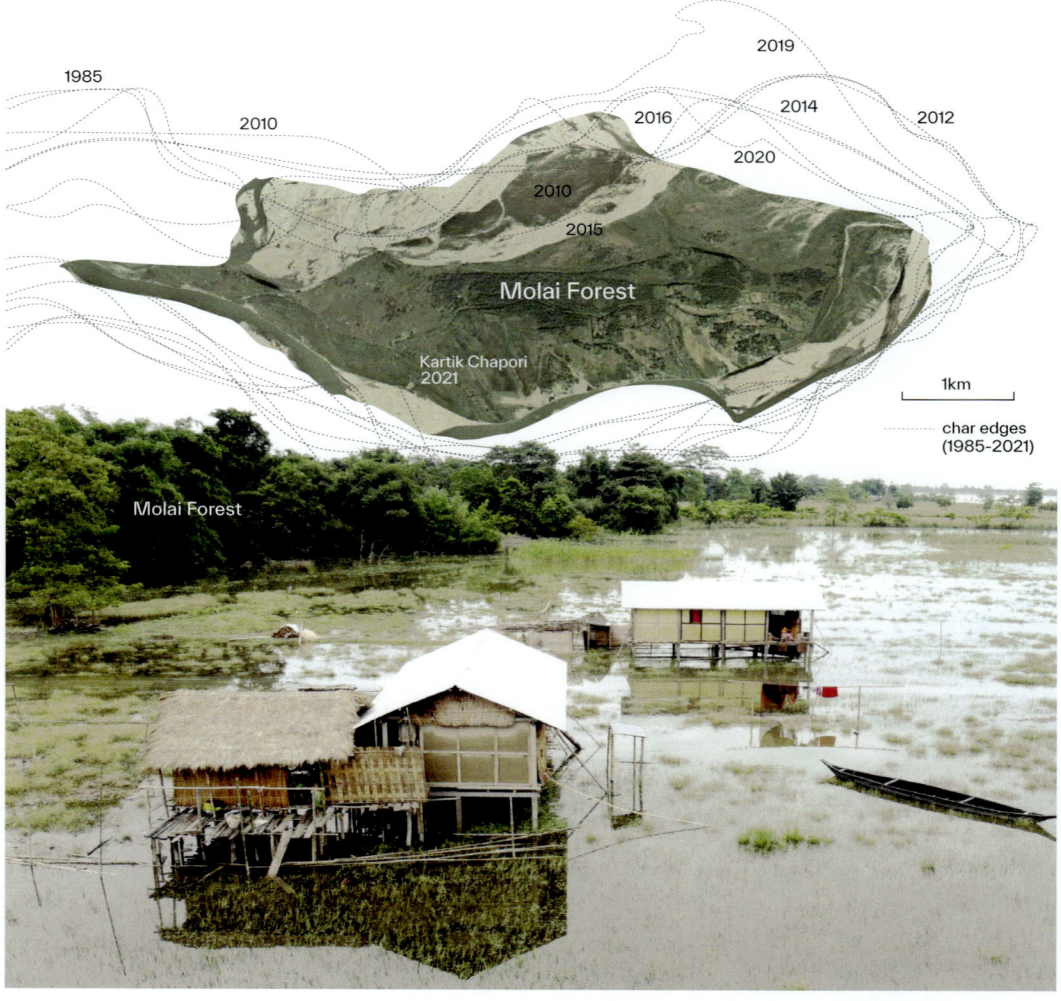

Living with/in the fluid landscape. Mising stilt houses are crafted to adapt to fluctuating water levels, evident in the dotted lines representing the shifting land/water edges during significant flood years, constantly reshaping the territory.

Foresting a *char*

Against the backdrop of the emergence of social forestry in India during the 1970s, the social forestry wing of the Assam State Forest Department initiated a bold experiment in 1979. The objective was to assess the effectiveness of tree cover in mitigating flooding and erosion, with a particular focus on *char-chaporis*, which were considered fallow lands by the state. An ambitious afforestation project aimed to cover a 200-hectare expanse within a sandbar, planning to plant thousands of trees over a five-year period (Bhattacharya 2018). However, due to a lack of funds, the project was soon abandoned. As the initiative was phased out, unpaid and angry workers abandoned these plantations.

The demise of the planting scheme faced an unexpected turn due to the persistence of Jadav Payeng, one of the forest workers, who continued planting new trees. Residing at the southern edge of the *char*, Payeng, along with fellow villagers, redefined the fate of the project. Over time, parts of the sandbar transformed into a flourishing forest.

> The forest was created through the knowledge we, Mising people, have about ecology. That is why I call it a people's forest. We used indigenous methods of soil preparation, planting seeds, and cow dung as manure to improve soil fertility, drip irrigation method to water saplings and releasing earthworms to prepare porous arable soil. I ferried red ants on my canoe to the *char*. (Payeng 2021)

Payeng's recounting highlights the pivotal role of traditional Mising ecological wisdom in transforming an abandoned project into a thriving forest. His assertion of the forest as a "people's forest" underscores the application of Indigenous knowledge in ecological stewardship, showcasing the adaptability and resourcefulness of the Mising community.

Popularly known as Molai Kathoni (*kathoni* means woodland in Assamese) after the nickname of Jadav Molai Payeng, today the forest spans 1,360 acres (5.5 square kilometres, hosting at least three rhinoceroses and more than four Bengal tigers, while witnessing a yearly migration of 150 Asian elephants (Bhattacharya 2018)). It provides not just wildlife habitat, but also ensures livelihood and food security for the local community by providing food, firewood, fodder and medicinal plants. Studies on the plant diversity and carbon stock of the human-created forest and on a natural forest of a comparable age showed that the plant species composition, plant diversity and carbon stocks of the former were similar to those in the latter (Guha 2021).

Despite being on government land, the local community shares collective responsibility for forest management. Payeng adheres to the Mising tradition of honouring nature by visiting the forest every morning, although this practice has become challenging in recent times due to increased man-animal conflicts. Until the 2000s, the forest's existence remained unknown outside the community. In 2008, the forest gained accidental recognition when a herd of elephants, tracked by state authorities, revealed a dense and expanding forest. As the forest gained global attention, reports of poaching for rhino horns and elephant tusks emerged. The lack of security in Molai Kathoni made it vulnerable to poachers.

As the forest began to gain more and more media coverage, Jadav Payeng came to be known by the state title of "Forest Man of India". Today, this case has become a symbolic landscape of environmental activism featured in newspapers, documentaries and children's books. Previously in Assam, Indigenous people residing on the fringes of forests were considered encroachers, and their historical sociocultural connection with nature was disregarded. In 1974, the forest department displaced Mising villages located within the UNESCO World Heritage site (i.e., Kaziranga National Park, situated partly in Golaghat District and partly in Nagaon District of Assam). The villages were then relocated to the periphery of the protected area (Cremin 2011). However, with international media focussing on Molai Forest, the nature of state involvement became scrutinized and led to the consideration of forests as community reserves. While the plantation programme can be viewed as a state effort to control barren hinterlands, nature, with a nudge from the community, charted its own course. This narrative suggests the futile attempts to control the uncontrollable: the *char*-scape and its inhabitants.

Inhabiting a *char*

The Misings, known for their resilient riverine lifestyle and adaptability to floods, have teetered on the edge of becoming climate refugees due to repeated flooding and erosion of their territories. An intriguing lens to view the Misings is through Scott's study

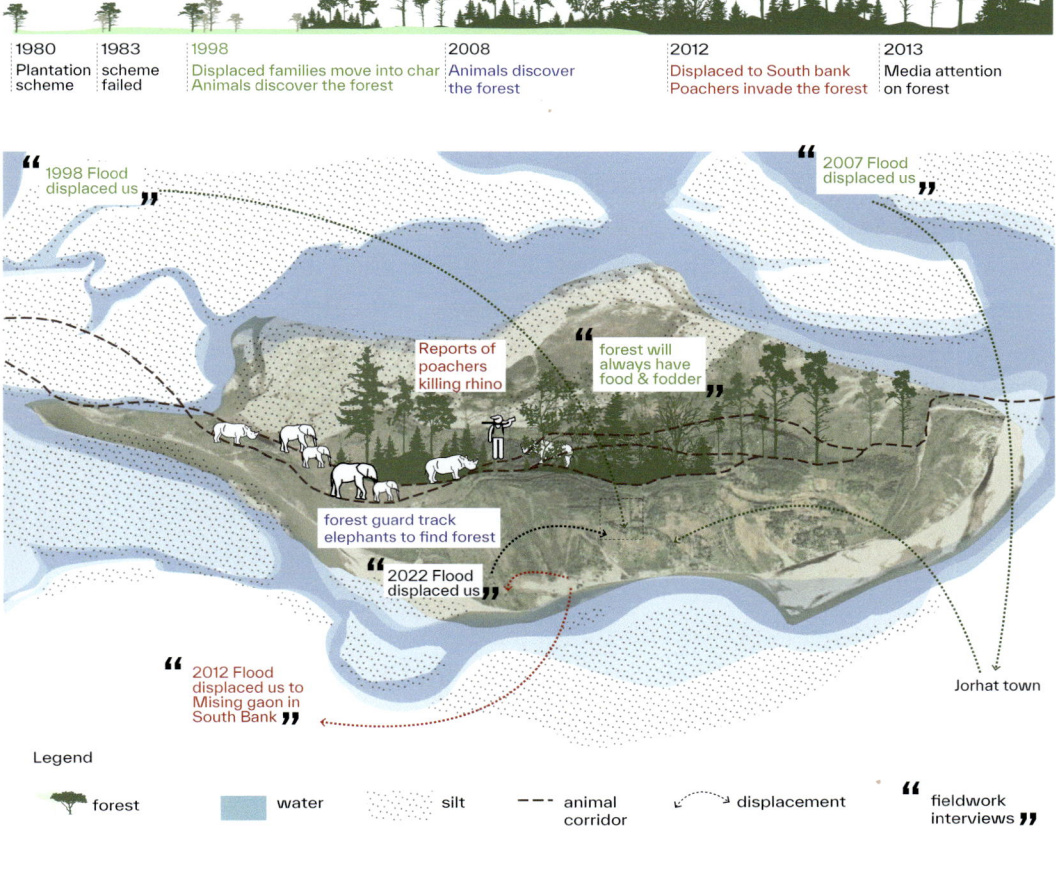

Foresting a *char*. The process of planting a forest on a shifting *char* resulted in an unforeseen transformation.

of "Zomia and hill societies primarily peopled by runaways from state-making projects in the valleys" (Scott 2009: 127). The Misings, who migrated to the plains from the hills of Arunachal Pradesh, regarded land as a common good. The boundaries of their territory were flexible, relying on access to natural resources. Once in Assam, in the face of erosion and accretion of sand deposits, their territories tend to be recomposed and redefined according to hydro-geomorphological dynamics.

Amidst the channels of the river, the eroded Mising territories have become sandbars and administratively obtained the status of public land. As occupants without rights or titles on public lands, they face the risk of expulsion. Communities losing their lands redefine their territories within or outside the framework of the territorial administration. Families displaced by the river continue their mobile way of life by occupying vacant land or land allocated by the territorial administration to establish their villages, without any formal rights or titles.

Fieldwork interviews revealed that the village of Kartik Chapori, initially located near the north bank, relocated to the studied *char* after the 1998 flooding, and since then locals refer to the char as Kartik Chapori. In 2006-2007, major flooding washed away several villages in the Upper Brahmaputra Valley. Village relocation plans were negotiated by Congress party members with territorial authorities of Jorhat District. State authorities proposed land on the south bank, near Jorhat town, to 42 families from a Majuli River island near the north bank. The new land proved unsuitable for rice cultivation and grazing, forcing the families to seek new livelihoods. In 2010, part of the village relocated to Kartik Chapori. The village was not officially registered on cadastral maps; it was a hybrid space, between land and water, with an uncertain status. The occupants first brought their cattle, and later they began growing rice and vegetables that they sell in the Jorhat market.

The floods of 2012 eroded the Mising village on the southern edge of the *char*. This land loss prompted many, including Payeng and his family, to move to the south bank of the Brahmaputra. Despite the relocation, community ties and connections to the forest endure to this day. Fieldwork conducted during the 2023 monsoon season revealed a new cluster of stilt houses next to the forest. The residents of the village who chose to stay after the 2012 flood relocated nearer to the forest following the erosion of 2022.

The migrations and displacements underscore how the forest's presence influenced community resettlement and economic activities in the *char*. Today around 217 families reside in Kartik Chapori. The discovery of the forest and settlements around it led to its incorporation into the official records as Kartik Chapori NC (i.e., non-cadastral village, implying resettled or rehabilitated

Settling with/next to nature. The transition – from the forest to grazing lands, then to a *haor* (a wetland ecosystem that floods during monsoons and retains water through the dry season), further to the Mising kitchen garden, and finally leading to the Mising settlements – forms intricate zones marked by their complexity and supporting diverse habitats.

villages). The "NC" designation signifies that the land is awaiting an official survey and no revenue map exists (Das 2022). This classification serves the revenue department as a strategy to address the challenges of mapping shifting territories while affording flexibility for inhabitants to move around.

Settling next to forest

Having once lived in the forests next to the hills of Arunachal, the Misings have adapted their knowledge to plant and sustain a forest in Kartik Chapori. Families, predominantly belonging to the Mising community, established *okum* (stilt houses) adjacent to the forested areas. In addition to the Misings, Kartik Chapori is inhabited by people from the Bihari community. They were brought by British officials from the state of Bihar to construct boats and ships (Guha 2021) and ended up staying in Assam.

The people living around Molai Kathoni are primarily livestock farmers, and their sustenance is intricately linked to the conservation of the forest and its surroundings, as well as the availability of resources around them. With threats like erosion and elephant raids impacting crop yields, many families have shifted to a reliance on milk trade as their main livelihood, accentuating their dependence on the forest's resources. The Mising lifestyle integrates with the forest ecosystem, blurring boundaries between humans and non-humans. Kitchen gardens, serving as extensions of the forest, provide essential produce during winter months, supplementing the subsistence farming (fig. 5). Despite prohibitions on timber collection, fallen trees from the forest's periphery are utilized for various purposes, illustrating an interdependence between human settlement and the forest.

The Misings are worshippers of *Do:nyi* (sun) and *Po:lo* (moon) and claim to be the children of nature (Pegu and Gogoi 1997). The Mising community's customary rituals, intertwined with forest resources, emphasize their spiritual connection to nature. The forest remains a repository of medicinal plants vital to Mising healing practices, and the community relies on forest-derived remedies due to limited access to medical facilities in the remote *char*. The long-term association with the forest, cultural acceptability and religious beliefs are crucial to the role of *mibu* (Mising shamans), who emphasize the coexistence of *uie* (spirits) with humans rather than dwelling in a separate realm. In addition, the making of *apong* (a traditional beverage of Misings) is dependent on many medicinal plants from the forest and is linked to their cultural identity. This interplay between the forest, humans and animals is crucial to the livelihoods of settlers around Molai Kathoni, as well as conserving the biodiversity of the area. Genevieve Michon points out how – aside from sustaining local livelihoods on a day-to-day basis and securing access to and use of local resources through locally defined and collectively accepted rules – community-managed forests embody shared property, tradition and memory, as well as intergenerational solidarity and interdependence (Michon et al. 2012).

Role of forest in Mising society. The forest, serving as an extension of Mising kitchen garden and grazing ground, intricately intertwines with local livelihood practices.

Towards a nuanced understanding of the forest/society relationship

Concerning the development (mainly infrastructural) and colonization of Indigenous territories, the Brahmaputra River has garnered immense interest in recent years. Rapid urbanization and threats of global warming call for innovative modes of habitation within the vulnerable ecologies of the Brahmaputra Valley which emphasize practices of collaboration with nature while accommodating diverse forms of knowledge and worldviews. Discussing multiple domains and types of knowledge, with differing logics, Agrawal highlights how "the same knowledge can be classified one way or the other depending on the interest it serves, the purposes for which it is harnessed, or the manner in which it is generated" (Agrawal 1995:

31). The case exemplifies how the same knowledge, directed by specific interests and generated through distinct approaches, led to the creation of a flourishing ecosystem, challenging conventional control efforts over the *char* territory and showcasing the adaptability and resourcefulness of the Mising community. What started as an experimental project to analyze the impact of tree cover against flooding and erosion is ambivalent today, situating itself between SEK (scientific ecological knowledge) and neo-Indigenous TEK (traditional ecological knowledge) approaches of the Mising community, blurring the boundaries of Indigenous and scientific knowledge. Evidently, the case does not provide direct answers for climate-adaptive spatial planning, but it does push towards the need to fundamentally reconceptualize the nature/culture relationship.

The idea of a pristine, wild and people-free nature to be conserved, propagated by the colonial imaginary and adopted by the Assam Forest Department, stands in stark contrast to Molai Kathoni – a people's forest. The Mising way of inhabiting and their relationship with nature sees the forest as a structure of multiplicity which accommodates settlements in a symbiotic way. Flexibility of resource use and environmental knowledge are factors for settling, as demonstrated by the Misings in Kartik Chapori. The case forwards human and animal communities that essentially learn to live together as the most effective modes of existence, highlighting non-humans left to live independently and in accordance to their own laws of cohabitation (Descola 2014). The continuity between humans and non-humans influences how the forest is managed.

The ways of settling and Mising forest stewardship are a manifestation of the communities' worldviews and the symbolism they attach to nature. The type of forest urbanism that the Misings demonstrate in Kartik Chapori offers a promising example of how spaces of multiplicity can merge, erase distinctions between nature and culture, and guide a new urbanism – a forest urbanism that challenges the supposedly scientific but often reductionist and utilitarian approach to forestry and urban planning. Community engagement and spiritual connection to nature through intergenerational values about revered forests led to the creation of a socially shaped landscape characterized by ecological balance.

Understanding the case from that perspective will present promising opportunities for the design of alternative strategies for settling with/in nature. This understanding can contribute significantly to resilience thinking of socio-ecological systems and forge newer forest-human (and more than human) alliances.

Acknowledgement

This research is funded by the Interfaculty Council for Development Co-operation (IRO) Doctoral Scholarships at the KU Leuven. Das also expresses gratitude to the Mising community of Kartik Chapori, whose openness and hospitality helped facilitating fieldwork in the *char*.

References

Agrawal, A. (1995). Dismantling the Divide Between Indigenous and Scientific Knowledge. *Development and Change*, 26, 3, 413–439. Available at: https://doi.org/10.1111/j.1467-7660.1995.tb00560.x.

Barbosa, L.C. (2015). *Guardians of the Brazilian Amazon rainforest: environmental organizations and development*. London, New York: Routledge, Taylor & Francis Group; Earthscan from Routledge.

Bhattacharya, B.K. (2018). 'Mulai Kathoni: A young man's dream is now a thriving "people's forest"', Mongabay. Available at: https://india.mongabay.com/2018/05/mulai-kathoni-a-young-mans-dream-is-now-a-thriving-peoples-forest/. Accessed December 21, 2022.

Cremin, E. (2011). Between land erosion and land eviction: Emerging social movements in the Mishing tribe fringe village of the Kaziranga National Park (Assam, Northeast India). In Dutta, D. and Saikia, J.R. (eds.), *Environment and development: Emerging issues and debates* (pp. 168–185). Guwahati: Planet Ink.

Das, B. (2022). In quest for a 'better' future, non-cadastral areas near Guwahati becoming ghost villages. Available at: https://thenewsmill.com/2022/07/quest-for-better-future-non-cadastral-areas-guwahati-kolongpur-ghost-villages/, https://thenewsmill.com/2022/07/quest-for-better-future-non-cadastral-areas-guwahati-kolongpur-ghost-villages/. Accessed March 3, 2024.

Descola, P. (2014). Configuration of Continuity. In *Beyond Nature and Culture* (pp. 33–38). Chicago: The University of Chicago Press.

Gadgil, M., & Guha, R. (1995). *Ecology and equity: The use and abuse of nature in contemporary India*. London and New York: Routledge.

Guha, N. (2021). Assam's man-made forest acts as natural hedge against floods. Available at: https://thefederal.com/features/jadav-payeng/ Accessed January 9, 2024.

Hunter, W.W. (1885). *The Imperial Gazetteer of India*. Available at: http://archive.org/details/in.ernet.dli.2015.11609. Accessed 9 January 9, 2024.

Jain, S. (1984). Women and People's Ecological Movement: A Case Study of Women's Role in the Chipko Movement in Uttar Pradesh. *Economic and Political Weekly*, 19, 41. Available at: https://www.jstor.org/stable/4373670?casa_token=doo39SNt3z4AAAAA%3AUFBZSrhZ53sAcLSl4es0uICXAhAcL8dRwz90AE57n_d9OxWzbhAw8_dBt8CGAAKNmqsP3vAXSn7YWON2utzVf-2fRE4PSQ0NKSl8eh8-RgKlfG61Tw5Kg.

Lahiri, S.K., & Sinha, R. (2012). Tectonic controls on the morphodynamics of the Brahmaputra River system in the upper Assam valley, India. *Geomorphology*, 169–170, 74–85. https://doi.org/10.1016/j.geomorph.2012.04.012.

Michon, G. (2005). *Domesticating forests: how farmers manage forest resources*. Paris Jakarta (Indonésie) Nairobi (Kenya): CIFOR ICRAF.

Michon, G., et al. (2012). Forests as Patrimonies? From Theory to Tangible Processes at Various Scales. *Ecology and Society*, 17, 3, 1–10. https://doi.org/10.5751/ES-04896-170307.

Payeng, J. (2021). Interviewed by Swagata Das. January 6[, 2021], Kokilamukh, Assam.

Pegu, D., & Gogoi, D.J. (2021). Religion of Mising Tribes in Assam. An Overview. *Remarking an Analisation*, 5, 12 (March, 2021), E-21–E-27.

Scott, J.C. (2009). *The art of not being governed: An anarchist history of upland Southeast Asia*. New Haven, London: Yale University Press (Yale agrarian studies series).

Scott, J.C. (1998). *Seeing like a state: how certain schemes to improve the human condition have failed*. Veritas paperback edition. New Haven, CT London: Yale University Press (Yale agrarian studies).

Sharma, C.K., & Sarma, I. (2014). Issues of Conservation and Livelihood in a Forest Village of Assam. *International Journal of Rural Management*, 10, 1, 47–68. Available at: https://doi.org/10.1177/0973005214526502.

Thakur, N. (2021) The struggles of a 'River People' in Assam. *Sapiens*. Available at: https://www.sapiens.org/culture/mising-river-people-assam-india/. Accessed December 2, 2023.

III. VISIBLE AND INVISIBLE FORESTS.
THE CULTIVATION OF SHADE IN WINNIPEG, MANITOBA, CANADA

Kamni Gill

Cities are constituted by the visible forest: the structure, spacing and species of trees. Shaped by physical conditions, the visible urban forest creates an arboreal interplay of rooms, passages and clearings that define a city within a wider region by providing continuities or contrasts with natural or agricultural systems. The woodlands, groves, avenues and glades of a community can be equivalent to streets, buildings, squares – building blocks of recombinant urbanism which define rich spaces for human inhabitation. Their structure, spacing and pattern create conditions of enclosure, variety, intricacy and rhythm. Trees constitute an ambiguous space in between the enclosure of buildings and the openness of landscape, a space characterized by a change in air and light: it is cooler, moister and darker under the canopy of trees. Trees give urban public space a spatial and sensorial coherence.

The trees of the urban forest also perform a range of measurable ecological functions. There has been a proliferation of research methods that demonstrate how trees can contribute to a climate-resilient city through their effect on urban cooling and stormwater attenuation (Pataki 2021). As much as trees contribute to the form of a city, they also contribute to its processes. Trees are instruments of urban design and ecosystem services. Other studies quantify the substantive impact trees have on well-being, correlating a robust urban forest to healthy citizens (Pataki 2021). Yet this effect is not simply physical. The urban forest is a complex aesthetic construction whose visible forms and management practices represent dimensions of urban tree planting which are sometimes not easily perceived. How, where and if trees are planted and by whom expresses a particular set of ideas of what it means to live in a particular place at a particular time. Trees can define places of magic, of ceremony, of production, of ornament, of habitat, of ecology, of utility or form part of a sentient breathing, living network. Nor are such understandings of trees and the spaces they define fixed. The same trees can provoke multiple, invisible cultural associations.

How can the infrastructural functions of the urban forests and the critical role they play in climate change adaptation be made more visible in climate change planting initiatives through an attentiveness to the spatial experience of treed spaces and to their cultural implications of urban trees in Winnipeg, Manitoba, Canada?

The state of the forest and the state of the climate in Winnipeg

Winnipeg is located between the extensive boreal forest of the Canadian Shield to the East and the Aspen Parkland and Prairie to the West. The typical hottest summer day was 34.4 between 1976 and 2005 and this measure is expected to increase to 39.5 between 2051 and 2080. During the same period, the typical coldest winter day was -35.9, which is expected to increase to -28. The winter is long and cold, while the summer is short and hot. Conditions of drought are common and intensive spring flooding characterize the region (Climateatlas.ca). Weather extremes make establishing trees as a cooling green infrastructure difficult: the number of species that can be planted is limited and the rate of growth slow (Diamondhead Consulting Ltd (DCL) 2021: 31).

Winnipeg's tree canopy of 17% (DCL 2021: 8) consists of a limited range of tree planting types: woodland, avenue trees, park land and residential gardens. The distribution of these four tree types corresponds to urban morphologies. Riparian forests provide connective recreational corridors along sections of Winnipeg's rivers and streams. Aspen-Oak woodlands characterize larger scale urban forests. Such trees are under the jurisdiction of the City of Winnipeg's Naturalization Services. A combination of avenue trees and privately planted trees creates a canopy cover of about 80% in older residential neighbourhoods. In newer suburbs, there is some provision of street trees, though the canopy cover is approximately 10% due to the age of trees and a less consistent spacing, less spontaneous regeneration of trees and fewer trees in residential gardens (DCL 2021: 6). The large-scale creation of parks and street tree plantings was most prevalent in the 1900s, where treed public spaces were conceived to reflect the philosophical considerations of the City Beautiful movement and of landscape architects such as Frederick Law Olmsted (MacDonald 2014: 24–32). Trees were a natural counter to the city, providing places to appreciate nature. Street and parkland trees now comprise about 10% of the total tree canopy and number 300,000 (DCL 2021: 9), creating an important matrix of shade in the neighbourhoods where people live and walk. They are managed by the Urban Forestry Branch.

These treed residential areas are fragmented by treeless transportation infrastructure: railways, highways, electrical rights-of-way and large-scale commercial/light industrial zones. The heavily built-up commercial centre is dominated by impervious surfaces and single trees. Surface parking is the dominant public open space downtown and includes only sporadic tree planting. The systematic removal of trees from Winnipeg streets began in 1904, when elm trees planted along sidewalks were removed in favour of vehicular access (MacDonald 1995: 16). Winnipeg, with its limited range of species and tree planting types, is also limited in its shade.

Climate change and urban forestry practices

Winnipeg's *State of the Forest Report* (DCL 2021: 8) quantifies the ecological contribution the current canopy makes to mitigating the effects of climate change through an i-Tree analysis. The report notes that Winnipeg's 3,000,000 urban trees comprise seven genera: elm, ash, linden, maple, spruce, oak and poplar, with elm and ash together comprising 24% of the total tree canopy and 58% of street and parkland trees. Of these, elm contributes most significantly to carbon capture, leaf surface area (shade provision) and stormwater attenuation (DCL 2021: 8).

The climate change resilience of the public tree canopy is threatened by Dutch elm disease and emerald ash borer. The Urban Forestry Branch focuses its efforts on the control of these two diseases in the street and parkland trees that fall under its jurisdiction. The Forestry Branch removes about 9,000 publicly owned trees a year. Nearly 40% of these removals are the result of disease, but these trees are replaced at an average of only 2,000 per year. There is already a deficit of 14,500 trees and up to 40,600 vacant planting sites before considering an increase in canopy cover as a climate change strategy and with a straight 1:1 replacement ratio of mature trees lost with new trees with a calliper of 50-70 mm (DCL 2021: v).

A demanding climate and a car-centric urban morphology make new tree establishment difficult. The existing tree matrix provides a framework for an urbanism of shade, but there is not sufficient density or distribution of other tree types or planting techniques to support movement protected by tree shade between neighbourhoods. Finally, despite a general climate change policy that recognizes the necessity of the urban trees, the planting and stewardship of the urban forest is funded at a level of only $12.5 million a year, which is insufficient to address comprehensive disease management, the growing tree replacement deficit and maintenance of young trees (DCL 2022: 45). Nor can local nurseries provide the supply of trees needed to maintain the current canopy or provide sufficient trees to expand planting efforts, according to the city's arborist.

Considering the collective

Winnipeg has significant strengths in responding to climate change initiatives. Local activist groups such as TreesPlease! (treespleasewinnipeg.com) have argued for increasing urban forestry and recognizing trees as urban infrastructure. Trees Winnipeg (treeswinnipeg.org), another prominent local tree group, supports a million-tree initiative that has planted 29,000 trees since its inception in 2019. Climate change tree planting initiatives emphasize replacing individual trees or reaching a particular number of trees planted or a percent canopy cover. The spatial impact of these efforts in Winnipeg does not always reflect the infrastructural importance of tree planting efforts nor their contribution to climate change adaptation strategies. Trees themselves could communicate new conditions for climate change resilience in Winnipeg through an infrastructural scale planting that also works as strategic urban design. Aligning tree planting to specific urban morphologies beyond the residential street or parkland creates new types of urban spaces that reinforce the relationship between the city, its trees and future climate resilience. Rather than supporting new construction within the City's riparian corridors, the extension of the city's riparian forest could be prioritized for the contribution that rivers in combination with trees make to urban cooling and biodiversity (Cheng 2019).

Instead of replacing trees along high speed transportation corridors where salt, exhaust and snowplows jeopardize tree survival, broad swathes of lawn along bike path verges could be treed at forest densities. Trees would protect routes from sun, wind and snow accumulation offering increased thermal comfort and an appealing alternative to cars.

Collective planting prioritizes the survival of trees and the provision of shade through composition of an arboreal interplay of routes, enclosures and clearings. Such an approach allows for a climate-responsive urbanism that uses trees themselves as the modest media for demonstrating alternative ways to occupy urban spaces – ones that value trees and human movement in shade over vehicular movement. As trees become more integral to urban living, so can the forms and techniques of planting reflect their importance as a necessary infrastructure. The impact of planting trees collectively is recognized by landscape architects such as Henry Arnold, who argues that trees are gregarious and grow naturally as communities, and that effective spatial design requires trees that are similar to forest-grown forms, planted close together to create a continuous canopy of foliage (Arnold 1980: 47–48). Michel Desvigne gives tree volumes the same weight as built infrastructure through forest-like plantings in the Parc aux Angeliques in Bordeaux "…where the density and forms applied [to trees] allow for a multitude of situations where everyone can find a place" (Desvigne 2020: 86-87). The value of considering the trees as a collective is important, also for strategically identifying where trees should most effectively be planted to support climate adaptation measures in relation to human communities.

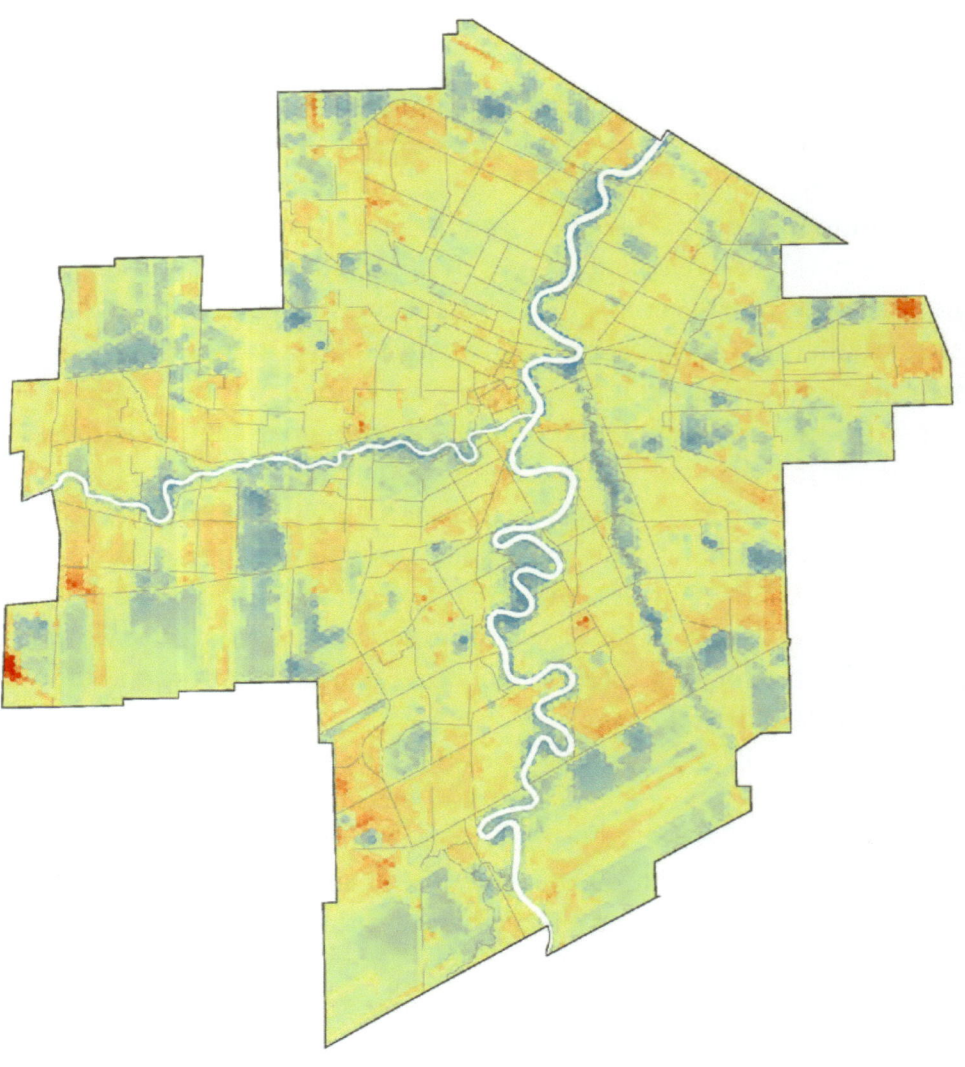

The urban heat island effect in Winnipeg. The presence of riparian trees extends the cooling function of Winnipeg's two rivers to adjacent streets. Blue indicates the cooler areas.

Planting trees as collectives also enhances ecological processes. Trees are networked organisms that protect and nurture each other (Simard 2020: 4). Considering modes of planting that acknowledge trees as part of a continuous matrix of soil, water, air and other organisms, rather than as discrete objects, results in a wider range of tree species, spaces and techniques of tree planting. Keeping existing tracts of woodland, allowing for spontaneous regeneration, planting densely spaced whips and planting more trees at smaller callipers are all recognized modes of establishing

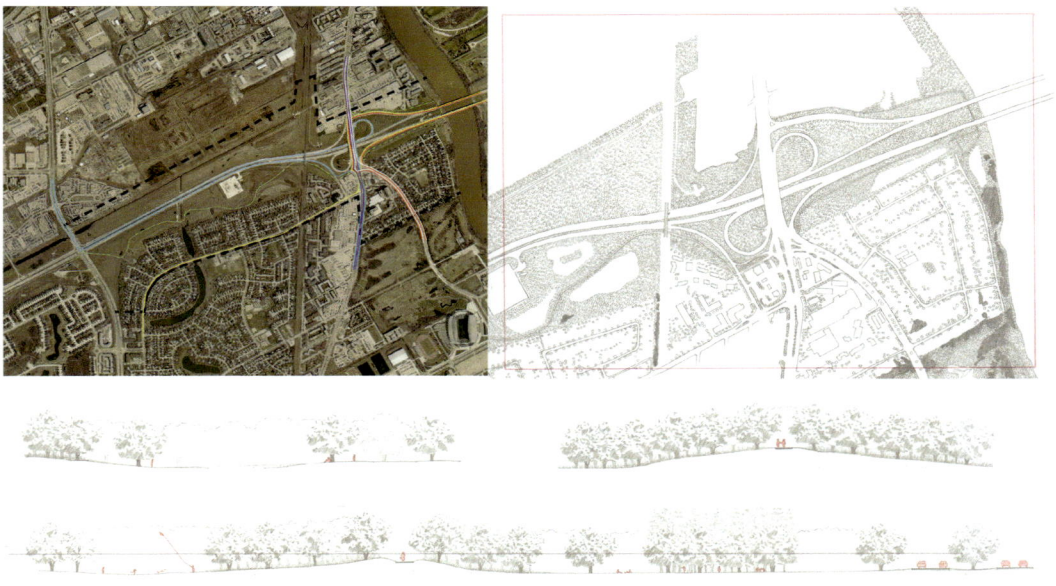

Afforestation of verges supports bicycle use, shaded walking routes to school and a series of community glades. The intensive use of trees gives spatial and sensorial coherence to land left vacant from the creation of road infrastructure.

trees that could offset more expensive tree replacement measures in light of the exigencies of Winnipeg's climate and the limited budget. Using two-tier establishment techniques in which trees provide shelter to each other could foster tree growth. Planting smaller trees at higher densities as a vegetative community in urban settings creates a network of relationships between other trees, soil, air and organisms, which results in greater tree health and survival and provides viable alternatives to conventional street tree planting.

An infrastructural scale cultivation of tree types which creates correspondences between site morphologies and an expanded range of planting techniques is not new. Just as the City of Winnipeg now needs to expand the urban canopy, so did farmers in early twentieth-century Manitoba need to expand rural tree planting types beyond the provision of shade for houses through shelterbelts or the provision of fuel through woodlots. An enlarged lexicon of tree planting types for soil preservation, water conservation, biodiversity, snow protection, maintenance of roads and ditches, fruit production and recreation set the conditions for a significant

Spontaneous regeneration of an Aspen woodland along an existing bike path in Winnipeg. Higher densities ensure tree coverage over time.

An agricultural forest matrix composed of a wide variety of distinctive tree types. Dominion nurseries in Brandon, Manitoba supplied trees and information on tree planting strategies that corresponded to particular soils, topographies, microclimates and functions that were critical to one kind of settlement on the Canadian Prairies.

expansion of an agricultural forest matrix on the Prairies. Between 1901 and 1935, the establishment of nurseries and federally funded experimental farms supplied 145 million trees designed for prairie hardiness. They provided farmers with new planting types that outlined a range of establishment techniques – keeping existing copses, allowing natural regeneration on deforested lands, transplanting whips, planting specimens – and their utility in relation to the exigencies of local weather, soils and hydrology (Ellis 1945).

Collective to culture

Creating new tree types on the Canadian prairie was necessary for agricultural settlement – an inexpensive way of responding to local climate. Trees became an agent of culture, expressive of one way of surviving in a place. As new forms and techniques of planting the urban forest are determined, it is important to acknowledge the multiple cultural dimensions that could inform tree planting efforts. The Winnipeg Urban Forest Strategy recognizes the need to acknowledge the city's trees as both a cultural and an ecological resource (DCL 2021: 2; DCL 2022: 6, 39). However, contemporary urban and historical agricultural afforestation has tended to reflect a narrative tied to land ownership by usually European settlers, ignoring Indigenous uses of local trees.

The creation of the Red River settlement at the site of Winnipeg in the late 1800s led to nearly complete deforestation of riverbanks and the loss of any rights to trees for Indigenous people. However, the first cultural manifestation of trees in the region is in the use of the Red River's riparian woodlands as winter camps by Indigenous people. Aspen and oak bluffs provided winter shelter, fuel and food. Trees were the source of medicine, berries, and maple syrup. Wood was harvested using parts of trees that could regenerate such as poles for tepees or bark for scrolls (Morgan 2020: 42ff). The current designation of Winnipeg's spontaneously generating woodlands as 'natural' ecosystems erases historical Indigenous inhabitation of the riparian forest prior to European settlement and contemporary uses. The Red River and its trees still hold cultural significance for Winnipeg's Indigenous communities. They shelter

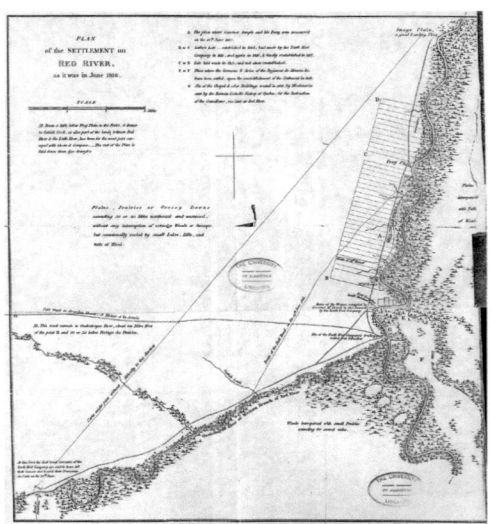 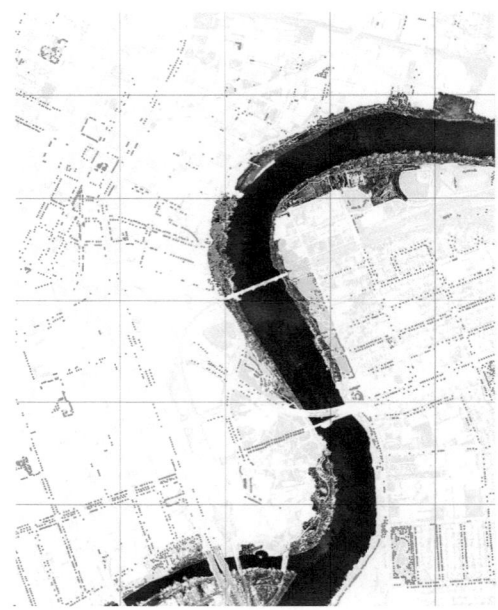

The configuration of trees demarcates different forms of settlement over time. Dense riparian woodlands punctuated with clearings for encampments characterized Indigenous settlements in the 1800s – a contiguous, communal treed space. In the twenty-first century, while public thoroughfares are lined with trees and the riparian edge remains, the urban forest is fractured by private treed gardens and paved, treeless commercial areas.

homeless encampments. The itinerant population of the city is 60-70% Indigenous, and homeless people continue to rely on the riparian forest for shelter and fuel. Temporary memorials for missing and murdered Indigenous women and lost youths are placed within the city's riparian woodlands, and they still serve as place of ceremony. What is officially considered a natural, ecological system is also a productive cultural space of inhabitation and ritual.

  Urban parks and woodlands become places for acknowledging Indigenous histories of displacement and a necessary Indigenous cultural resurgence through public art. Though trees often provide settings for Indigenous cultural artefacts, they are not themselves considered cultural, expressive of an alternative way of relating to a place. The woodlands of pre-settlement endure in the Winnipeg landscape, with original birch-aspen stands extant in the landscape, but these can be readily cleared as if they were not integral either to the cultural life of Winnipeg's citizens or to the urban ecosystem. Woodlands are not systematically integrated

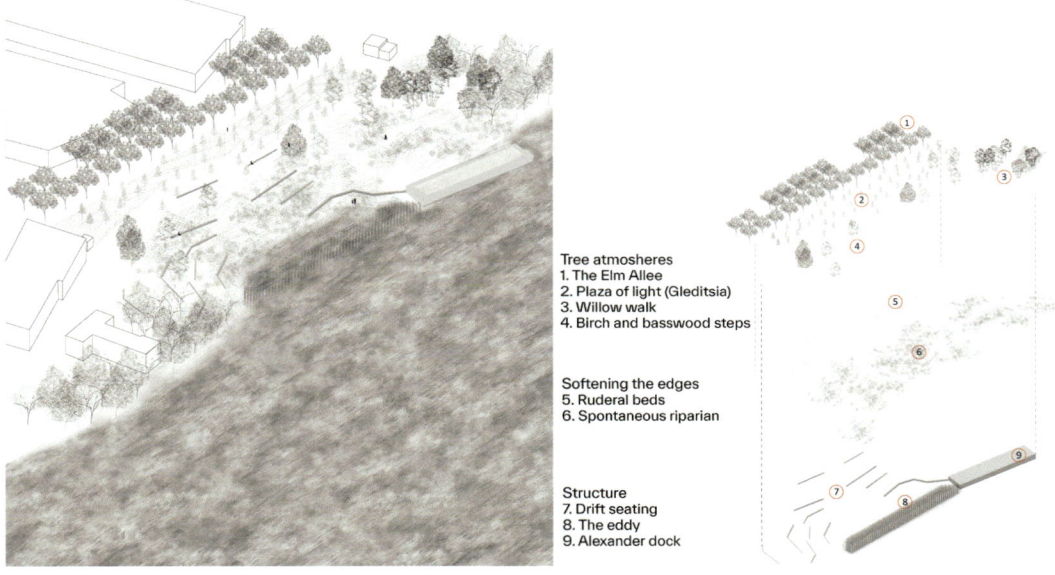

Drift garden. The drift garden cedes the shaping of the land and the dissemination of seeds to the river and natural processes of succession. Trees create a series of distinctive spaces through the careful composition of shade and clearings, thereby bringing the city to the edge of the river and the river and its trees into the city. Tree types include an allée of elm, a grove of honey locust, a willow walk and spontaneously regenerating steps of basswood and birch.

into urban settings. In urban tree policy, they are still emblematic of 'wilderness', natural spaces apart from human settlement—despite their cultural significance. Boulevards, parking lots, playgrounds and forest remnants among other spaces could be enriched by more actively acknowledging how a wider diversity of people relate to urban trees. This position gives a cultural dimension to the ecological and spatial affordances of dense tree planting described earlier.

In a final studio project at the University of Manitoba, the designer shapes a series of forest gardens. He retains an existing memorial to missing and murdered Indigenous women, strengthening its contemplative character through the modulation of shadow by trees. He explores a relational understanding of water and trees drawing on traditional Indigenous knowledge. He sets up a connection to the river under the canopy of trees through a "drift garden" that accepts of what is already there, rather than controlling the natural processes of flood, wind and forest regeneration. The drift garden is "tended by the river" whose floods and mud seed a volunteer basswood and birch forest (McPherson 2020). The space of trees and their establishment express a different way of dwelling under the canopy. It is an interpretation of what it means to recover and restore Indigenous rights to the trees of the city through the medium of the trees themselves – a different conception of the relationship of urban nature to culture.

The agency of trees and the agency of people

Using trees to express alternative cultural relationships through their poetics enriches urban spaces and entwines the visible qualities of trees with the invisible understanding of how trees have different associations for different people. Rethinking the city grove by grove as spaces that are expressive of diverse cultural ways of relating to nature strengthens the role trees play in quotidian urban life and their importance to Winnipeg's liveability as temperatures rise.

The City of Winnipeg is a fertile ground for cultural, spatial and ecological experimentation with trees at the strategic and the site-specific scale, especially considering the climate change pressures it faces. Winnipeg, with large tracts of open space and a temperature that is several degrees higher than its agricultural environs, could support an intensive network of experimental field stations dedicated to new urban tree cultures.

A strategy that focuses on the culture of tree spaces rather than the replacement of individual specimens offers a regenerative approach to the use of trees. It recalls a recommendation by one early Winnipeg settler that each citizen of the new city should be entitled to least five hectares of wooded commons to ensure their survival in a place racked by climate extremes (Shay 2015). Expanding the ways people can participate in urban afforestation offers a way of creating a diversity of relationships between public urban spaces and trees; of cultivating alternative conditions of shade through a richly configured, extended vocabulary of tree spaces that give both people and trees agency in the making of a climate-resilient city.

References

Arnold, H. (1980). *Trees in Urban Design*. New Haven: Van Nostrand Reinhold.

Atuhairwe, Calvin. (2019) Sylvan City Studio, University of Manitoba. Unpublished student work.

Cheng, L., Guan, D., Zhou, L., Zhao, Z., Zhou, J. (2019). Urban Cooling Island Effect of Main River on a Landscape Scale in Chongqing, China. *Sustainable Cities and Society*, 47, 101501. https://doi.org/10.1016/j.scs.2019.101501.

Diamondhead Consulting Ltd (DCL). (2021). *State of the Urban Forest Report*. Winnipeg: City of Winnipeg.

Diamondhead Consulting Ltd (DCL). (2022). *Winnipeg Draft Comprehensive Urban Forest Strategy*. Winnipeg: City of Winnipeg.

Ellis, J. H. (1945). *Farm Forestry and Tree Culture Projects for the Non-Forested Region of Manitoba. A Report Prepared for the Post-War Reconstruction Committee of the Government of Manitoba*. Winnipeg [: Manitoba Advisory Committee on Woodlots and Shelter-belts].

Imbert, D., Tiberghien, G. A., & Faigenbaum, P. (2020). *Transforming landscapes: Michel Desvigne paysagiste*. Berlin: De Gruyter. https://doi.org/10.1515/9783035609974.

MacDonald, C. (1995). *A City at Leisure: An Illustrated History of Parks and Recreation Services in Winnipeg 1893-1993*. Winnipeg: City of Winnipeg, Parks and Recreation Department.

MacDonald, C. (2014). *Making a Place: A History of Landscape Architects and Landscape Architecture in Manitoba*. Winnipeg: Manitoba Association of Landscape Architects.

McPherson, E. (2021). *Fluid Relations: Reframing water on the edge of the Red River* [Master of Landscape Architecture Practicum]. Winnipeg: University of Manitoba.

Morgan, R. G. (2020). *Beaver Bison Horse: The Traditional Knowledge and Ecology of the Northern Great Plains*. Regina, Saskatchewan: University of Regina Press.

Pataki, Diane E., Alberti, Marina, Cadenasso, Mary L., Felson, Alexander J., McDonnell, Mark J., Pincetl, Stephanie, Pouyat, Richard V., Setälä, Heikki, Whitlow, Thomas H. (2021). The Benefits and Limits of Urban Tree Planting for Environmental and Human Health. *Frontiers in Ecology and Evolution*, 9 (April), 1–9, 603757. doi: 10.3389/fevo.2021.603757.

Shay, T. (2015). Pioneers on the Forest Fringe: The Wood Economy of the Red River Settlement, 1812-1813. *Manitoba History,* 78 (Summer, 2015), 2+.

Simard, S. (2012). *Finding the Mother Tree: Discovering the Wisdom of the Forest*. Toronto: Allen Lane Canada.

IV. BRIDGING GREEN GAPS. EMPOWERING PARTICIPATORY GOVERNANCE THROUGH TREE PLANTING IN BARRANQUILLA, COLOMBIA

Alejandra Parra-Ortiz, Gina Serrano-Aragundi

Urban forestry research has shown that green infrastructure distribution, in particular tree canopy cover, tends to reflect and maintain socio-economic disparities in cities since low-income neighbourhoods often have less vegetation than affluent ones (Schwarz et al. 2015; Cruz-Sandoval et al. 2020). This tendency exacerbates the vulnerability of lower-income populations to global warming due to limited access to the benefits of green infrastructure, such as temperature regulation and stormwater filtration. There is a need for transformational change in governance to address this environmental inequality, since existing public policies can exacerbate problems when power dynamics are not considered in their formulation and implementation (National Academies of Sciences, Engineering, and Medicine 2021).

Since Colombia ranks as one of the most unequal countries in the world in terms of wealth and income (Chancel et al. 2021), it is no surprise that there are great disparities in green infrastructure and ecosystem services distribution in the nation's four biggest cities: Bogota, Medellin, Cali and Barranquilla (Escobedo et al. 2015; Rubiano Calderón 2019; Herrera-Hurtado 2019; Serrano-Aragundi 2020; Shiraishi 2022). However, reforesting cities evenly is challenging due to underlying factors such as limited resources and technical capabilities, as well as migration tendencies that shape the urbanization of extremely dense informal settlements. An important component to address inequities is the incorporation of more community engagement in public policies. Participation aids in balancing top-down power dynamics that perpetuate inequalities, since human nature leads people to defend strategies they helped create regarding their daily life improvement (Fischer 2012; Dawes et al. 2018).

Barranquilla has received international recognition for its urban forest management (Tree Cities of the World 2019; 2020). It is also Colombia's first city to commit to the transformation of its urban development model towards a "biodivercity" in harmony with nature (Khatri et al. 2022). Barranquilla Verde, the city's environmental authority, has developed a socially inclusive tree planting programme following a three-step methodology which includes the identification of priority areas to reforest, engagement of the community in the re-planting, and follow-up with the community to ensure the long-term survival of the trees planted.

Focus on Barranquilla

Barranquilla, located on the western margin of the Magdalena River Delta, is the main economic centre in the Caribbean region of Colombia and the fourth most populated city in the country (1,310,163 people in 2022, according to Gobernación del Atlántico, 2023). It has a dry tropical climate with temperatures ranging between 21.4°C and 33.3°C. The rainy period is from May to November, with 60-173 mm/month. The dry period runs from December to April and is characterized by strong northeast trade winds (CIOH 2010).

In Colombia, all neighbourhoods are classified into six socio-economic strata; the first three are the most economically disadvantaged and their services are meant to be financed by the top two strata, five and six; the fourth stratum is neutral, and people must only pay for their own public services (Congreso de Colombia 1994). In Barranquilla, neighbourhoods are grouped within five localities: Riomar, Norte-Centro Histórico, Suroccidente, Suroriente and Metropolitana (Alcaldía de Barranquilla 2020). Barranquilla Verde has full jurisdiction over the urban areas of the city while the Regional Autonomous Corporation manages its remaining rural areas. As an autonomous entity, Barranquilla Verde is responsible for all the environmental administrative functions and service provision of the territory; it coordinates efforts with the municipality and complies with the Ministry of Environment and Sustainable Development guidelines. To complement the Mayor's Office goal

Barranquilla is a bustling seaport city flanked by the Magdalena River and the Atlantic Ocean.

of planting 250,000 trees in public areas, Barranquilla Verde initiated its tree-planting initiative in June 2021. This effort, targeting non-public areas, is ongoing with no specified end date, aiming to plant more than 1,000 trees per year in gardens, backyards and patios, as well as inside private and public institutions.

The tree planting programme is financed through compensation for tree cutting licences granted by Barranquilla Verde; licenced users must provide and plant the trees or give the economic equivalent to Barranquilla Verde. The tree compensation rate varies from 1:4 to 1:5 depending on tree maturity, origin and status conservation. Initially, the programme was stocked with eight thousand trees paid in instalments by one company, the first instalment was in 2021, and 500 trees were planted by December 2022. The pilot programme required few technical skills and resources, since its financing is assured through Colombia's tree compensation laws.

Tree canopy cover analysis and planning strategy

The tree canopy cover was assessed for neighbourhoods in four localities: Norte-Centro Histórico, Suroccidente, Suroriente and Metropolitana. Riomar was excluded since it has the city's highest canopy cover (Serrano-Aragundi 2020) and its neighbourhoods range between the top three socio-economic strata. Google Maps aerial imagery from the municipality's archives were sourced as geographic information system (GIS) files outlining each neighbourhood's boundaries. The files were then loaded into the open source software i-Tree Canopy v. 7.1, which assigns random points in the images for the user to manually classify each point as "tree" or "non-tree". After categorizing 300 points, the standard error stabilizes below 5%, allowing accurate quantification of tree canopy cover percentages across all analyzed neighbourhoods.

The data were classified into eight sorts of tree canopy cover percentages, ranging from 5-10% to >40%, separated in 5-point increments (5-10%, 10-15%, etc.). Each class was assigned a shade of green to visualize the canopy cover of each neighbourhood on a map in QGIS (fig. 3). Neighbourhoods below 20%

canopy cover were designated high priority for tree planting, as this level is below the recommended minimum vegetation cover of 30% (Konijnendijk 2021). A linear regression analysis was carried out to test the correlation between the tree canopy cover and socio-economic stratification per neighbourhood, the former (y-axis) as the dependent variable and the latter (x-axis) as the independent variable. The analysis was conducted using the RStudio statistical software version 2022.07.2 and the *lm* function, which provides the P-Value, Multiple R-squared and adjusted R-squared. These values indicate the strength and significance of the correlation. A list was created to describe the 84 tree species suited for planting according to the ecology of the city, while also prioritizing native and tropical fruit tree species in order to strengthen urban biodiversity, contribute to food security and incentivize the cultural identity and culinary traditions linked to trees. The list included detailed information like leaf type and conservation status. Finally, a one-page questionnaire was developed to register the programme's beneficiaries and gather information about the existing trees on their properties, their individual knowledge about trees, their needs and conflicts regarding trees with their neighbours, and so on. To promote openness, the beneficiaries were interviewed in a conversational manner when filling out the questionnaire, allowing the staff to confirm if the conditions were clear and to better assess their likelihood of giving proper maintenance to the trees; the process also allowed beneficiaries to express the programme's limitations.

Through the mayor's offices, Barranquilla Verde contacted councillors (localities' elected representatives), presidents of Community Action Boards (neighbourhoods' elected representatives), public schools, prisons and the Early Childhood Program, the last being a public initiative to ensure children's rights and health in Barranquilla. Local security forces and private companies were contacted independently. In-person visits to each household were possible through elected representatives and by approaching beneficiaries of the urban agriculture programme of Barranquilla Verde, aimed towards reducing food insecurity in poorer neighbourhoods. The programme beneficiaries revealed available planting areas and expressed their preferred tree species; their wishes were considered

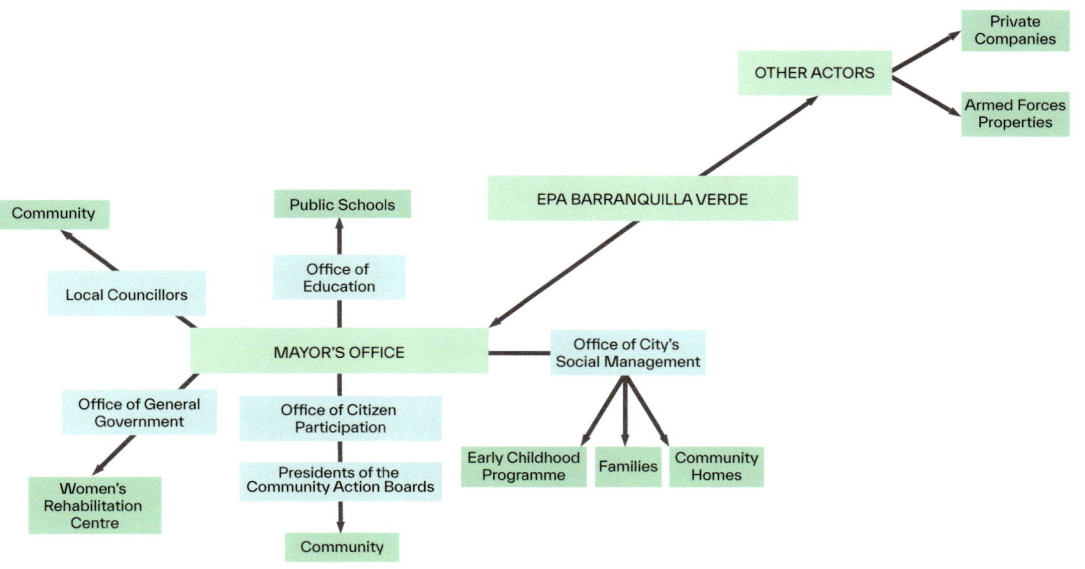

Numerous actors were involved during the first year of the tree planting programme. The easiest way to approach the community was partnering with other public programmes without political party affiliations.

in relation to tree species availability and technical determinants that would affect tree survival, like soil quality. Lastly, a volunteer partnership was agreed upon with the army, the civil defence group and local environmental NGOs to assist with tree planting.

In follow-up with the communities, small informative posters of tree species and participation certificates were handed out to beneficiaries together with a signed memorandum of their responsibility to report about the trees through in-person visits by Barranquilla Verde staff, phone calls, email or WhatsApp messages. Following Colombia's legal standards for tree compensation, trees must be monitored every six months for two years after planting to ensure their health and survival; if a tree dies within that period, it must be replaced. Continued communications provided insight into the challenges of leaving the full responsibility for tree maintenance to the beneficiaries and, through a partnership with the Early Childhood Program, beneficiaries with minors that planted fruit trees continued receiving assistance to combat food insecurity and malnutrition.

An EPA Barranquilla Verde member interviewing a prospective participant for the BAQ Cultiva programme. (June, 2021)

Tree canopy cover analysis results

In total, 155 neighbourhoods were analyzed: 22 in Metropolitana, 31 in Suroriente, 41 in Norte-Centro Histórico, and 61 in Suroccidente. Metropolitana and Suroriente had, respectively, the highest (22.7%) and lowest (16.7%.) tree canopy cover. Only ten neighbourhoods exceeded 30% tree canopy cover (in Suroccidente and Norte-Centro Historico), eight of which were located in the western basin, a protected area prone to landslides and where urban development is limited. The linear regression results provided a p-value of 0.04, showing a relationship between the tree canopy cover and socio-economic strata with 96% of confidence interval.

Neighbourhoods with less than 20% of tree canopy cover were prioritized and accounted for almost 70% of the total. One out of ten neighbourhoods had less than 10% of tree canopy cover, which is far from enough to adapt to climate change. Seventy-one households registered during the first six months of the programme, impacting 335 people in low strata neighbourhoods with low tree canopy cover. During tree plantings, people initially reluctant to participate showed more openness towards the staff, indicating that the programme could increase its impact as trust

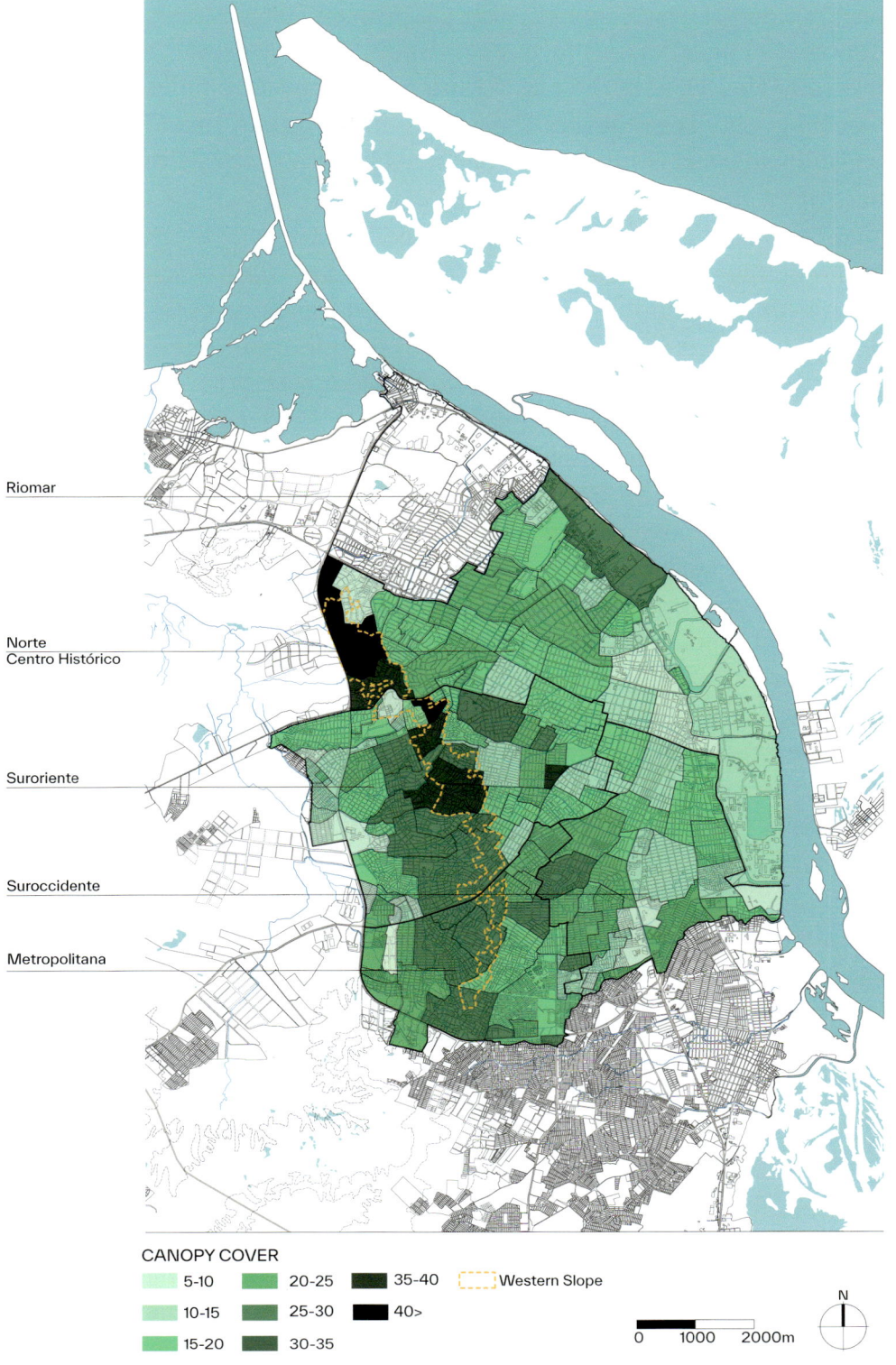

Tree canopy cover by neighbourhood.

from potential beneficiaries increases. Among the biggest impediments to having trees in their backyards was the need to build a new or bigger house, either for their children or to rent. In addition, they also mentioned conflicts between neighbours caused by the ownership of trees, in particular due to plant litter and underground breakages caused by tree roots. The biggest impediment to maintaining the trees was the limited access or high cost of water. Tree survival was also threatened by soil quality since several homes were built upon construction debris mixed with the soil.

Available trees were limited to commercial species, involving few options of native and fruit trees despite their great acceptance by the communities. Since Barranquilla Verde does not have a plant nursery, the entity depended on producers who did not align with the objectives of the programme. The most demanded tree species were those bearing fruit commonly found in the local diet: guanabana (*Annona muricata*), followed by guava (*Psidium guajava*), and sapodilla (*Manilkara zapota*). Working in partnership with the Early Childhood Program ensured that families receiving fruit species were already aware of their importance for a balanced diet. Interestingly, older household members often mentioned having fruit trees in their backyards during their childhood and reported that they presently ate less diversity of fresh fruits and traditional fruit desserts than before.

The interviews showed that those who immigrated to Barranquilla from elsewhere in Colombia participated in the programme as a way to improve their living conditions. Many new migrants were fleeing the internal armed conflict. Adults of all ages signed up, while women registered almost three times more than men. Finally, the most common pests found were termites and the parasite plant *pajarita* (family *Loranthaceae*). Captive birds and turtles (including endangered species) were the most common pets in households after dogs and cats. Some beneficiaries practised animal and vegetable farming. After six months, the survival rate of trees planted ranged from 71% to 92 % per neighbourhood. Among the reasons behind the tree mortality are domestic animals, insufficient soil quality and damage by machinery. Only 8% of trees presented pests, for which organic recommendations were provided.

Lessons distilled from the quantitative study

The correlation between low tree canopy cover and socio-economic strata in Barranquilla is in line with previous studies in Colombia (Escobedo et al. 2015; Rubiano Calderón 2019; Herrera-Hurtado 2019; Serrano-Aragundi 2020; Shiraishi 2022). The i-Tree Canopy tool proved to be practical for directing urban planning in Colombia's context since tree canopy cover results were clear, unbiased, uncomplicated to obtain and easy to compare with socio-economic strata data. However, uncertainty associated with random point sampling and remote sensing methods must be considered (Parmehr et al. 2016) and could account for differences between this study and Barranquilla's previous one using the same method (Serrano-Aragundi 2020). To reduce errors, sampling 500 to 1000 points – depending on the size of the area – is suggested (Parmehr et al. 2016). It should be noted that this size sample would double or triple the time to complete the same task, and exceptional precision is not always necessary; public entities only want to know where to direct their efforts in the most efficient way possible.

Previous studies in Cali and Bogota provided insights for understanding why green inequalities exist in relation to the impact of forced migration from rural areas to major cities due to Colombia's armed conflict (Escobedo et al. 2015; Rubiano Calderón 2019; Shiraishi 2022). Public policies have been inefficient in properly addressing the incorporation of new settlers; consequently, many high-density and spontaneous settlements have been constructed in peri-urban areas, leaving little space for parks and other green areas (Ardila et al. 2006; Parra Delgadillo 2018). Other scholars have pointed out that Latin American states must coordinate the resolution of social inequalities with environmental programmes since major social vulnerability tends to equal major vulnerability to climate change (Santelices & Rojas, 2016; ECLAC 2018). Therefore, Barranquilla's sustainable transition towards a "biodivercity" must integrate the participation of communities that have become accustomed to not being considered in policies that directly affect them, as well as a transition out of poverty and social discrimination for marginalized communities.

A household member in company with EPA Barranquilla Verde members and military volunteers during a planting day in October 2021.

Such a tree planting programme allows policy-makers to observe the interactions between environmental and social problems. It could also be the entry point for nuanced socio-ecological thinking (see Biggs et al. 2021) towards governance, an approach that aligns with the "biodivercity" concept in understanding the interconnection between nature and society. Furthermore, the tree planting strategy resulted from the collaboration between academia, policy-makers and the community, a triangle of stakeholders that builds knowledge from very different perspectives, generating robust results. The input from the academia allowed for deeper reflection that otherwise would not have happened between policy-makers and the community, since municipal employees are too busy trying to generate impact strategies within limited timelines and the community is not able to closely observe the dynamics inside public offices, only the results.

On the other hand, communications about the tree planting programme often mentioned its importance to fight global warming. However, when other factors are considered like water usage and burnt fossil fuels during logistics, the net carbon sequestration might not be considerable in these types of projects (Pincetl et al. 2013). This is particularly true for this programme since the funding originates from compensation for tree cutting. Improper communications about the programme and lack of transparency could easily lead to greenwashing. This planting programme does not have a net-positive climate impact. It is firstly a legal obligation to compensate ecological damage and, secondly, is a strategy to mitigate green inequalities. Parallel policies to reduce tree cutting should also be developed.

Follow-up with the community

The successful tree survival rate during the first six months of the programme was most likely influenced by the tree maintenance provided by the beneficiaries. Many of them had experience gardening and growing food from Barranquilla Verde's urban agriculture programme, which taught several families how to grow plants from seed. A primary limitation of the programme was that Barranquilla Verde ought to have either its own plant nursery or enough access to trees fitting with the goals of the programme. These troubles were compensated by collaboration with the urban agricultural programme and using the seeds from fruits that come in the food baskets of the Early Childhood Program Eventually, a full chain collaboration could incorporate hard-to-find tree species.

Final recommendations for the follow-up include incorporating discussion groups in the neighbourhoods to promote exchange of experiences and prevent or solve common issues related to trees. The moderate use of technology could be useful, keeping in mind that plenty of people in poorer neighbourhoods have limited access to the Internet; a simplified version of the questionnaire should be made available through an official automated WhatsApp account in order to reach more of the prioritzed communities. On the other hand, domestic disturbances that generate permanent damage to the trees must be counteracted before or during planting by securing the trees with barriers. Watering was the major limiting factor of tree maintenance, such that techniques like adding hydrogel near the root of the trees when being planted and setting drip irrigation systems are two options worth exploring; creating a circle with the soil around the tree to trap and direct more water over the roots and mulching are other common techniques useful for the local weather conditions that should be included in the starter kit that comes with the tree.

Bridging green gaps through governance

The results show the need for a transformational change in local public policies that tackle social inequalities together with environmental programmes. There are preventive and remedial public policy measures that need to be redirected to improve tree cover in vulnerable sectors. First, city planners must include predictions of national and foreign immigration in urban development projects. Participatory governance can better assure a safe and appropriate establishment of new settlements to avoid informal constructions. All informal settlements located in areas prone to natural disasters must be relocated to safer zones that offer public services and green infrastructure, again, through participatory governance. More research is necessary to assess a sustainable transition of current informal neighbourhoods to organized blocks with increased green areas.

There is a need to develop further studies to foresee the possible correlations between different tree canopy distributions along the variety of socio-economic strata. The tool i-Tree Canopy requires little technical skills to get accurate measures of urban tree canopy cover, as well as to differentiate between other different types of land use (i.e., bare soil, grass, etc.). It has a fast-learning curve, and as such it can be used to guide public policies related to green infrastructure in cities, in particular to continue measuring the correlation between green areas and socio-economic strata as a means to select priority areas to reforest. On the other hand, compensation funds coming from permits of tree cutting can ensure the continuous funding of public planting projects, as long as more trees are planted than those being removed, since the survival rate of planted seedlings is never 100%, and because mature trees provide more ecosystem services than small trees. Future research should focus on

Participants during a planting day at the "Infancia de Primera" Early Childhood Development Center. (August, 2021).

the co-creation of nature-based solutions with the local population which promote planning space for trees when constructing new buildings, for the benefit of increasing the price of their investment and making their settlement more resilient to global warming.

Acknowledgement

The authors would like to thank Claudia Lucía Rojas Bernal for developing the Barranquilla maps.

References

Alcaldía de Barranquilla. (2020). 'Descubre – Conoce a Barranquilla – Localidades'. https://www.barranquilla.gov.co/descubre/conoce-a-barranquilla/territorio.

Ardila, G., Echeverri, C., and Universidad Nacional de Colombia, eds. (2006). *Colombia: Migraciones, Transnacionalismo y Desplazamiento*. Colección CES. Bogotá: Universidad Nacional de Colombia, Facultad de Ciencias Humanas, Centro de Estudios Sociales : Ministerio de Relaciones Exteriores : Fondo de Población de las Naciones Unidas.

Chancel, L., Piketty, T., Saez, E., Zucman, G. (2021). *World Inequality Report 2022*. World Inequality Lab.

CIOH. (2010). Climatología de Los Principales Puertos Del Caribe Colombiano: Barranquilla. 2010. https://www.cioh.org.co/meteorologia/Climatologia/ResumenBarranquilla2.php.

Congreso de Colombia. (1994). *Artículos 89 & 102 – Estratos y Metodología. Capítulo IV – Estratificación Socio Económica. Ley 142 de 1994 – Por La Cual Se Establece El Régimen de Los Servicios Públicos Domiciliarios y Se Dictan Otras Disposiciones*.

Cruz-Sandoval, M., Ortego, M.I., and Roca, E. (2020). Tree Ecosystem Services, for Everyone? A Compositional Analysis Approach to Assess the Distribution of Urban Trees as an Indicator of Environmental Justice. *Sustainability*, 12, 3, 1215. https://doi.org/10.3390/su12031215.

Dawes, L. C.,.Adams, A.E., Escobedo, F.J.,. Soto, J.R. (2018). Socioeconomic and Ecological Perceptions and Barriers to Urban Tree Distribution and Reforestation Programs. *Urban Ecosystems*, 21, 4, 657–671. https://doi.org/10.1007/s11252-018-0760-z.

ECLAC, United Nations. (2018). Summary (LC/SES.37/4). In *The Inefficiency of Inequality*. Santiago: Economic Commission for Latin America and the Caribbean (ECLAC).

Escobedo, F.J., Clerici, N., Staudhammer, C.L., and Tovar-Corzo, G. (2015). Socio-Ecological Dynamics and Inequality in Bogotá, Colombia's Public Urban Forests and Their Ecosystem Services. *Urban Forestry & Urban Greening*, 14, 4, 1040–1053. https://doi.org/10.1016/j.ufug.2015.09.011.

Fischer, F. (2012). Participatory Governance: From Theory to Practice. In Levi-Faur, David (ed.), *The Oxford Handbook of Governance* (pp. 457-471). Oxford: Oxford University Press. https://doi.org/10.1093/oxfordhb/9780199560530.013.0032.

Gobernación del Atlántico. (2023). Anuario Estadístico 2022. Capítulo 2: Demografía. Secretaría de Planeación https://www.atlantico.gov.co/index.php/anuarios-estadisticos/22501-anuario-estadistico-2022

Herrera-Hurtado, M.A. (2019). Morfología Urbana y Clima Local: Alternativas de Diseño Urbano a Partir de Infraestructura Verde – Caso de Estudio: Medellín, Colombia [M. Sc. in Environment and Development thesis]. Medellín, Colombia: Universidad Nacional de Colombia.

Khatri, A., Bustamante, D., Ruta, M., Garcia-Reyes, C., Thompson, F., Kohli, S., Pantelidou, H., Free, M., Magnani, G., & Schemel, S. (2022). *BiodiverCities by 2030: Transforming Cities' Relationship with Nature* (p. 51) [Rapport de synthèse]. Cologny/Geneva: World Economic Forum.

Konijnendijk, Cecil. (2021). The 3-30-300 Rule for Urban Forestry and Greener Cities. *Biophilic Cities Journal*, 4, 2.

National Academies of Sciences, Engineering, and Medicine. (2021). *Progress, Challenges, and Opportunities for Sustainability Science: Proceedings of a Workshop in Brief*. Edited by Paula Whitacre and Emi Kameyama. Washington, D.C.: National Academies Press. https://doi.org/10.17226/26104.

Parmehr, E. G., Amati, M., Taylor, E. J., & Livesley, S. J. (2016). Estimation of urban tree canopy cover using random point sampling and remote sensing methods. *Urban Forestry & Urban Greening*, 20, 160–171. https://doi.org/10.1016/j.ufug.2016.08.011

Parra-Delgadillo, J. (2018). Migraciones En Colombia (Ciudad-Campo): Análisis al Neorruralismo y Las Nuevas Ruralidades En Las Afueras de Bogotá (Cundinamarca). Bogotá, Colombia: Universidad Externado de Colombia.

Pincetl, S., Gillespie, T., Pataki, D.E., Saatchi, S., and Saphores, J. D. (2013). Urban Tree Planting Programs, Function or Fashion? Los Angeles and Urban Tree Planting Campaigns. *GeoJournal*, 78, 3, 475–493. https://doi.org/10.1007/s10708-012-9447-x.

Rojas-Bernal, C. L. (2023). Barranquilla: Ciudad y Paisaje 2023. Proyecto de investigación: Barranquilla: Atlas del Agua. Universidad del Norte.

Rubiano Calderón, K.D. (2019). Distribución de La Infraestructura Verde y Su Capacidad de Regulación Térmica En Bogotá, Colombia. *Colombia Forestal*, 22, 2, 83–100. https://doi.org/10.14483/2256201X.14304.

Santelices Spikin, A., & Rojas Hernández, J. (2016). Climate Change in Latin America: Inequality, Conflict, and Social Movements of Adaptation. *Latin American Perspectives*, 43, 4, 4–11. https://doi.org/10.1177/0094582X16644916.

Schwarz, K., Fragkias, M., Boone, C.G., Zhou, W., McHale, M., Grove, J.M., et al. (2015). Trees Grow on Money: Urban Tree Canopy Cover and Environmental Justice. *PLOS ONE*, 10, 4, e0122051. https://doi.org/10.1371/journal.pone.0122051.

Serrano-Aragundi, G. (2020). Diagnóstico Del Arbolado Urbano En El Distrito de Barranquilla En El 2019. Reporte sin publicar. Barranquilla, Colombia: Establecimiento Público Ambiental Barranquilla Verde.

Shiraishi, K. 2022. The Inequity of Distribution of Urban Forest and Ecosystem Services in Cali, Colombia. *Urban Forestry & Urban Greening*, 67 (January), 127446. https://doi.org/10.1016/j.ufug.2021.127446.

Tree Cities of the World. (2019). Tree Cities of the World. 2019 Summary.

Tree Cities of the World. (2020). Tree Cities of the World. 2020 Summary.

V. URBAN FORESTS AS POST-MANICURE OUTDOOR DESIGN TYPES

Jörg Rekittke

Residing in Berlin may indeed seem a privilege, given the city's remarkable abundance of parkland compared to other European counterparts. However, this numerical advantage fails to mask the stark reality of the city's green spaces. Professionally maintained trees and shrubs have become increasingly scarce, playground sand replacement has ceased, and public trees receive watering only through residents' voluntary efforts, often at their own expense. Sustaining urban parks over the long term necessitates qualified gardeners and robust management, rather than relying solely on well-intentioned volunteers. The once esteemed profession of urban gardener, emblematic of a city described by former Mayor Klaus Wowereit as "poor but sexy", now teeters on the brink of extinction. Professional gardeners have been supplanted by unskilled labor, leading to a reduction in maintenance activities to a bare minimum. The consequences of this neglect are readily apparent. Stefan Tidow, former state secretary in the Senate Department for the Environment, Transport, and Climate Protection in Berlin, lamented the deplorable condition of many parks in the city. Despite Berlin boasting over 2,500 public parks and green spaces covering a total of 1,000 hectares (Hofmann & Hönicke 2019), satisfaction among citizens remains notably low, falling well below the national average according to the German Garden Office Managers' Conference (Deutsche Gartenamtsleiterkonferenz 2022). Austerity measures implemented in post-reunification hit Berlin's green space offices (*Grünflächenämter*) particularly hard, resulting in a significant reduction in staff numbers (Senatsverwaltung für Finanzen 2022). Consequently, the maintenance of gardens and parks suffered, despite the city's proud reputation for its extensive greenery. The prevailing approach shifted towards prioritizing ease of maintenance, leading to a decline in demand for design elements such

Hausburgpark in the Friedrichshain district of eastern Berlin. Located in a hip neighbourhood within the urban railway ring, with a high proportion of middle-class property and townhouse stock, its poor condition represents absolute normality in Berlin.

as hedges, bushes and trees which require time and resources to upkeep. The Berlin chapter of the Association for the Environment and Nature Conservation Germany (BUND) has observed a troubling trend in the maintenance practices of external companies contracted by the Parks Department (BUND 2019). Rather than prioritizing careful maintenance, preservation and ecological enhancement, these contractors often resort to radical pruning or complete removal of vegetation, justified under the misleading guise of sustainability. This shortsighted approach disregards the long-term ecological integrity of urban green spaces in favour of short-term cost savings. The result is an unsustainable handling of these vital public areas, driven by budgetary constraints and staffing shortages. This failure to adequately care for Berlin's green spaces amounts to an ecological bankruptcy declaration, reflective of a broader trend towards minimalistic "savings parks: a bit of grass, a couple of sturdy metal benches, and that's it" (Hofmann & Hönicke 2019) – prioritizing basic amenities over ecological and aesthetic considerations.

Rubbish bins as park and forest rangers

In this chapter, our initial focus is on addressing local issues, figuratively cleaning up our own backyard before extrapolating potential lessons for other cities. Our decision is not solely based on our local ties but also inspired by the methodology of Alfred Russel Wallace (1823-1913), a revered British naturalist and explorer. Wallace's approach of commencing his scientific inquiries within the immediate vicinity of his lodgings resonates with our strategy (Wallace 2016). Until 2023, Berlin's governance was characterized by a 'red-red-green' coalition, comprising the Social Democratic Party of Germany (SPD), The Left, and The Greens. In 2020, Oliver Schruoffeneger, the district councillor for Urban Development, Building and Environment, emphasized the need to transform urban planning into a comprehensive urban development strategy, envisioning Berlin as a sustainable, socially conscious, and climate-neutral city of the future (Entwicklungsstadt Berlin 2020). Schruoffeneger highlighted the inadequacy of the allocated budget for park cleaning and staffing, revealing that the funds amounted to a mere six cents per square metre of green space. This stark figure underscores the depth of the challenges facing Berlin's park landscape and its users. A notable initiative emerged in 2015, spearheaded by Matthias Kollatz, then state minister of finance in Berlin, who enlisted the City Cleaning Service of Berlin (BSR) to alleviate the burden on district green space offices. The BSR, the largest municipal waste disposal company in Germany, gradually assumed responsibility for maintaining various green spaces, starting with the forest district of Teufelssee. By 2020, the BSR Park and Forest Service had been formally commissioned by the Berlin state government, overseeing maintenance in 79 green spaces and priority

Playground *Schleidenplatz* in the Friedrichshain district, East Berlin. The commissioned BSR ensures that the visitor sees bright orange rubbish bins from every angle.

areas across seventeen forest districts. The BSR professionals adopt a responsive approach to cleaning, adjusting their efforts based on weather conditions, visitor numbers and the level of waste accumulation (Berliner Stadtreinigung 2022). While this approach has proven effective in ensuring cleanliness and hygiene, some aesthetic concerns have been raised by users (Bartels 2018). The proliferation of bright orange rubbish bins, while functional, detracts from the visual appeal of the green spaces, presenting an unintended visual disruption. Bright orange rubbish bins are placed in such a large number that the visitors see several of them at the same time, no matter which way they turn their head. In summary, our exploration of Berlin's park maintenance strategies highlights both successes and aesthetic compromises, prompting reflection on the balance between functionality and visual harmony in urban green spaces.

Blankensteinpark in the Friedrichshain district, East Berlin.

       The overdose principle – a key component of Berlin's "Overall Strategy for a Clean City" formulated by the Berlin Senate for Economics, Energy, and Public Enterprises – has proven to be effective. In quintessential bureaucratic fashion, the underlying cause-and-effect relationship is articulated as follows: "Enhancement of the city's infrastructure to provide adequate waste disposal options and improved management of overflowing trash receptacles" (Hoffmann 2021). It is important to note, however, that this concept currently applies to only 79 sites, a fraction of the more than 2,500 public parks and green spaces throughout the city (Berliner Stadtreinigung 2022). The stark reality remains that Berlin's green space offices are severely under-equipped financially, a situation projected to persist in the long term and likely exacerbate the decline in the city's greenery (Hofmann & Hönicke 2019).

City forestation instead of city administration

The title alludes to one of the most iconic artworks of the 'documenta' exhibitions and one of Germany's most captivating artists in recent memory: Joseph Beuys. His masterpiece, *7000 Eichen – Stadtverwaldung statt Stadtverwaltung* (7000 oaks – city forestation instead of city administration), entailed the planting of trees accompanied by stones at 7000 locations across the city of Kassel, Germany. Initiated by Beuys in 1982 for 'documenta 7', the project was completed in 1987 for 'documenta 8'. Beuys imbued his works with multiple layers of meaning, inviting all to engage with them. Hence, we feel emboldened to interpret his vision of urban forestation. Beuys foresaw the potential for forestation to counter administrative stagnation. *7000 Eichen*, often described as a *soziale Plastik* (social sculpture), stands as the most consequential intervention in the outdoor installations of Kassel's 'documenta' exhibitions. Art scholar Harald Kimpel asserts, "No other artwork intervenes as profoundly and sustainably in the topographical and social fabric of the city, nor does any other commit to the ongoing care and appreciation of this gift to the citizens of Kassel as enduringly" (Kimpel 2022). Beuys himself articulated, "When envisioning a sculpture that encompasses not just physical material but also mental substance, I was compelled by the notion of social sculpture" (Stiftung 7000 Eichen 2022). In an interview concerning the project, Beuys expressed his aspiration to bridge the realms of nature and humanity within their everyday environments, stating, "I aim to increasingly immerse myself in the intersection of nature and human concerns in their workplaces. This endeavour represents innovation; it fosters a healing process for the myriad challenges we currently face. (…) That is my foremost objective" (Stiftung 7000 Eichen 2022).

      As an academic landscape architect, we find profound inspiration in Joseph Beuys' artwork and the slogan "City Forestation instead of City Administration". Beuys' vision, seemingly ahead of its time, now resonates strongly with contemporary developments unfolding not only in the German capital but also in other major metropolises. When city administrations are compelled to scale back investments in the costly and labour-intensive

Joseph Beuys at documenta 7, 1982, starting his art project *7000 Eichen – Stadtverwaldung statt Stadtverwaltung*. Initially, the work consisted of 7,000 basalt steles and was completed in 1987.

maintenance of parks, alternative forms of public space must be considered – spaces that may be less meticulously groomed yet equally vital for urban life. Drawing from Beuys' visionary perspective, we contemplate the following potential trajectories for the evolution of urban parks. When urban administrations find themselves unable to fulfil their obligation and social responsibility to provide citizens with well-maintained green spaces for uninterrupted enjoyment, they may turn to forms of greenery that have existed long before the meticulously designed park became the norm in urban landscapes. At the opposite end of the spectrum from the manicured park lies the natural forest, while its managed counterpart stands as its globally prevalent twin. Managed forests, developed and accessible, serve as havens for relaxation and recreation – much like traditional parks. However, while the historical park represents a luxury afforded by those with means, the managed forest operates as an economic entity, aiming to generate profit and material resources. From the perspective of many a financially strained city administration (*Stadtverwaltung*), the concept of profitable urban forestation (*Stadtverwaldung*) becomes not only tempting but also inherently logical. Christophe Girot's exploration of landscape architecture in his book *The Course of Landscape Architecture* provides valuable insight into the intricate relationship between natural forests and cultivated parks (Girot 2015). Girot's identification of two archetypes – the forest clearing and the walled garden – serves as a theoretical foundation for our thesis, albeit in a reversed order. The manicured urban park aligns more closely with the archetype of the clearing than with the dense forest. Should investments in maintenance wane, the cultural form of the clearing may regress into a wooded area, giving rise to an urban forest. While maintenance remains crucial to keep such forests accessible and usable for the urban public, the approach can shift from neglect to the provision of uplifting beauty. An exemplary case of this transformation is the Berlin Natur-Park Südgelände, situated in the Schöneberg district on the former Tempelhof railway yard. This 18-hectare park, once an open industrial area, underwent a process of near-complete overgrowth before being redeveloped into an urban forest, illustrating the potential for revitalization and adaptation of urban spaces in harmony with nature.

An Urban Forest Age

It would not be far-fetched to envision the future of Western urban landscapes evolving into an Urban Forest Age. Urban forests not only represent a flexible alternative to meticulously manicured designs, but also serve as critical assets in combating global warming by sequestering carbon and reducing energy costs (Konijnendijk & Shannon 2022). Moreover, urban forests foster increased biodiversity, accommodate diverse ecological design approaches, and provide urban dwellers with a healthy and multifaceted leisure environment. In regions devoid of human presence or minimal settlement, favourable outcomes are scarce. The ongoing struggle to preserve remote rainforests serves as a poignant example. However, in urban settings where progressive action is more feasible, opportunities abound. The demand for green spaces persists among urbanites, yet dwindling financial support for extensive park maintenance may precipitate the transition of parks into urban forests. Recognizing the evolving nature of both city parks and urban forests, we sought a precedent for a hybrid model, ideally situated close to home. Our search led us to the seamlessly integrated spaces of Treptower Park and Plänterwald, nestled along the banks of the River Spree in Alt-Treptow, within the Treptow-Köpenick district, south of central Berlin. Spanning 88 hectares, Treptower Park was established from 1876 to 1888 under the guidance of Gustav Meyer, the municipal garden director. It is one of the four nineteenth-century parks preserved in Berlin, alongside Volkspark Friedrichshain, Volkspark Humboldthain and Viktoriapark, providing local residents with ample recreational opportunities. Notably, Treptower Park originated from the reforestation of the Köllnische Heide heathlands (Wikipedia 2021), embodying typical Berlin park characteristics, along with its attendant challenges. Traversing the expansive Treptower Park from north to south eventually leads one into the densely wooded environs of Berlin's Plänterwald. Unlike a park, Plänterwald is a forest in its truest sense, representing a distinct ecosystem within the urban fabric.

In international silvicultural circles, the practice is known as "selection cutting", a harvesting technique designed to foster an uneven-aged or all-aged stand structure by selectively harvesting individual trees or small groups. This method is believed to confer ecological benefits, such as enhanced carbon sequestration, while ensuring a steady supply of marketable timber (Clarke et al. 2015). Despite its occasional resemblance to a jungle, a selection cutting operation represents a managed forest, overseen not by underpaid park attendants but by skilled and knowledgeable foresters. Beyond its intrinsic beauty, shade and cooling effect, the Berlin Plänterwald is accessible via public paths. However, it also serves a dual purpose as a productive forest, yielding wood for commercial sale. Historically, the Plänterwald accommodated another commercial endeavour during the German Democratic Republic era – the Spreepark. Established as the state-owned enterprise Operation Kulturpark (VEB Kulturpark) in 1969, it operated until its closure in 2002. In 2014, ownership transferred to the state-owned

Plänterwald in the Alt-Treptow district of Berlin, situated alongside the River Spree, south of central Berlin. It is a publicly accessible, managed selection cutting forest, not a park. Biodiversity is high, the possible uses are extremely diverse, and its beauty can be positively overwhelming.

Grün Berlin GmbH, with redevelopment efforts initiated in early 2016 (Busch et al. 2017). Presently, Grün Berlin is spearheading the transformation of the former Spreepark into a new iteration, envisioned as an art and culture park catering to tourists. While we have inspected the project details and peered through the fence, we harbour reservations regarding the current vision (Grün Berlin 2022) for the future Spreepark. The proposed imagery appears overly idealistic, with clichéd design representations exuding an air of naivety. The promotional rhetoric paints a picture of vibrant colours, fun and happiness, proclaiming, "With the new Spreepark, the capital is regaining a place that expands local recreation and leisure activities with the opportunity to experience and discover art and culture in public space" (Grün Berlin 2022). While we extend our best wishes for success to all involved, we cannot ignore the grim

Amusement park *Spreepark*, located in the Berlin Plänterwald. After its closure in 2002, it became a 'lost place'.

financial realities facing the city. Should this optimistic endeavour fall short, nature and the forest will swiftly reclaim what is rightfully theirs. Commercial ventures like the Spreepark are but fleeting phenomena in the grand scheme of geological and environmental history. Eventually, this forest clearing will once again be enveloped by nature's embrace.

    The origins of the selection cutting in Berlin's Plänterwald trace back to the aftermath of the Ice Age, when the landscape underwent significant transformations. Situated within the Warsaw-Berlin glacial valley, this area witnessed the resettlement of flora and fauna. Until the early twentieth century, extensive portions of the land remained unsuitable for development due to the high groundwater level. Between 1823 and 1840, the forested area was cleared, and faced with unprofitable management prospects, it underwent reforestation as the Plänterwald around 1873, coinciding with the establishment of Treptower Park (Busch, Geyler-von Bernus, Kahl 2017). In the envisioned Urban Forest

Currently, the *Grün Berlin* company plans and promotes a new version of the *Spreepark*.

Age, after the novelty of attractions like the former and current Spreepark fades into history and the site eventually reverts to its natural state, the management of the corresponding urban forest will no longer fall under the jurisdiction of well-intentioned park administrations, whether public or private. Instead, it will be overseen by an urban forestry office staffed with skilled foresters dedicated to both utilizing and conserving this magnificent urban forest. This forest, for once in its existence, will have the opportunity to embrace its true identity and essence, transitioning from a mere park into a thriving ecosystem under the stewardship of urban foresters.

References

Bartels, Gerrit. (2018). Die BSR und die Parks: Körbe für den Kollwitzplatz (The BSR and the parks: Baskets for the Kollwitzplatz). *Der Tagesspiegel*, August 11, 2018. https://www.tagesspiegel.de/kultur/die-bsr-und-die-parks-koerbe-fuer-den-kollwitzplatz/22902392.html.

Berliner Stadtreinigung. (2022). Reinigung von Parks und Forsten: Sauberes Grün für die Hauptstadt (Cleaning of parks and forests: Clean green for the capital). Accessed March 6, 2022. https://www.bsr.de/parkreinigung-23237.php.

Busch, Dora, Geyler-von Bernus, Monica, and Kahl, Birgit. (2017). *Geschichte des Spreeparks* (History of the Spreepark). Berlin: Grün Berlin GmbH.

Clarke, Nicholas, Gundersen, Per, Jönsson-Belyazid, Ulrika, Kjønaas, O. Janne, Perssond, Tryggve, Sigurdsson, Bjarni D., Stupak, Inge, and Vesterdal, Lars. (2015). Influence of different tree-harvesting intensities on forest soil carbon stocks in boreal and northern temperate forest ecosystems. *Forest Ecology and Management*, 351 (2015), 9–19.

Deutsche Gartenamtsleiterkonferenz (German Garden Office Managers' Conference). Accessed March 5, 2022. https://www.galk.de.

Entwicklungsstadt Berlin. (2020). Im Interview: Oliver Schruoffenegger, Bündnis 90/Die Grünen. Accessed March 2, 2022. https://entwicklungsstadt.de/im-interview-oliver-schruoffeneger-buendnis-90-die-gruenen/.

Girot, Christophe. (2015).*The Course of Landscape Architecture: A History of our Designs on the Natural World, from Prehistory to the Present*. Farnborough: Thames & Hudson.

Grün Berlin GmbH. (2020). Spreepark. Accessed March 10, 2022. https://www.spreepark.berlin/konzept-und-grundlagen/vision-ziele/.

Hoffmann, Kevin P. (2021). BSR übernimmt ab Mai Reinigung von 33 Parks und Plätzen in Berlin (BSR will take over the cleaning of 33 parks and squares in Berlin from May). *Der Tagesspiegel*, April 5, 2021. https://www.tagesspiegel.de/berlin/um-bezirke-zu-entlasten-bsr-uebernimmt-ab-mai-reinigung-von-33-parks-und-plaetzen-in-berlin/27048302.html.

Hofmann, Laura & Hönicke, Christian. (2019). Warum so viele Berliner Parks verkommen (Why so many Berlin parks are degenerating). *Der Tagesspiegel*, September 8, 2019.

Kimpel, Harald. (2022). Stiftung 7000 Eichen: Die Geschichte, Die Fakten. Accessed March 14, 2022. http://www.7000eichen.de/?id=38.

Konijnendijk, Cecil, & Shannon, Kelly. (2022). Call for papers: Urban Forests, Forest Urbanisms & Global Warming. Developing Greener, Cooler & more Resilient Cities. Conference, Leuven/Belgium. June 27-29, 2022.

Senatsverwaltung für Finanzen (Berlin). (2022). Personalstatistik. Accessed March 11, 2022. https://www.berlin.de/sen/finanzen/personal/personalstatistik/artikel.13543.php#headline_1_6.

Stiftung 7000 Eichen. (2022). 7000 Eichen. Accessed March 14, 2022. https://www.7000eichen.de/index.php?id=2.

Wallace, Alfred Russel. (2016). *The Malay Archipelago: The Land of the Orang Utan and the Bird of Paradise*. Oxford: Beaufoy (Stanfords Travel Classics).

Wikipedia. Treptower Park. Last modified October 30, 2021. https://de.wikipedia.org/wiki/Treptower_Park.

VI. A FENCE THAT GREW A FOREST.
A STRATEGY FOR A PARK AT PACHACAMAC ARCHAEOLOGICAL SANCTUARY

Takako Tajima

Urban forests are important infrastructures for adapting cities to the effects of climate change (Pataki et al. 2021), and many cities have embraced tree planting programmes to benefit public and environmental health (Pincetl et al. 2013; Nowak & Dwyer 2007). Tree planting, however, does not come without costs, and in many arid regions where water is scarce, the cost of irrigating the resultant landscapes bear too high a burden to sustain.

A 2014 study on water sources for the world's largest cities found that over three-quarters of their water – water for some 1.2 (± 0.05) billion people – come from surface water sources dependent on precipitation for replenishment, often from hundreds of kilometres away (McDonald et al. 2014). This overreliance on distant surface water sources imperils water security especially in arid climates. In some regions, anthropogenic global warming appears to increase the frequency of the combined instances of low precipitation and higher temperatures (Mann & Gleick 2015), exacerbating the already precarious balance between the supply and demand for fresh water.

Water for urban forestry and otherwise, however, can and should be obtained from a range of means outside conventional sources. In coastal areas with persistent fog, harvesting water from the air is relatively simple and inexpensive (Schmenauer & Cereceda 1991), and "just like an underground aquifer, the water is there to be utilized" (Schmenauer & Cereceda 1994: 91). In an ecosystem restoration project off the coast of southern California, a team lead by the United States Geological Survey (USGS) with funding from the National Park Service (NPS) collects water from coastal fog to re-establish woody vegetation lost to overgrazing. 'Fog fences' and 'fog hats' made from modest means serve as proxies for endemic vegetation and provide shelter and fog drip for seedlings until they have enough leaves to drip fog on their own (USGS Western Ecological Research Center 2021).

In 2019, Peru's 2021 Bicentennial Projects Initiative, the Ministry of Culture, and the Municipality of Lima launched a design competition for a new park along the perimeter of the site of the Pachacamac Archaeological Sanctuary. A project was conceptualized to investigate the possibility of cultivating an urban forest in one of the most arid landscapes in the world without irrigation from conventional water sources. Inspired by the work in California, the project developed a strategy that would transfer moisture from the air into the ground and catalyze a new ecological succession.

The Sanctuary

The Pachacamac Archaeological Sanctuary (hereinafter the "Sanctuary") is an archaeological site on the coast of central Peru approximately 40 kilometres southeast of Lima. Although much of the Peruvian coastline is desert, the Sanctuary flourishes due to its proximity to the Lurin River where it benefited from water and alluvial deposits from the Andes Mountain range.

The Pachacamac Park design competition was developed in response to the plan and was announced on February 11, 2019 by 2021: Bicentennial Projects, in association with Peru's Ministry of Culture and the Municipality of Lima.

False start

The first vision, naively and simplistically, embraced the possibility of replicating Tokyo's Eternal Forest. Begun in 1915, the Eternal Forest is a 150-year, four-stage succession plan to cultivate a forest from scratch. A little over a hundred years since its inception, the forest is developing as projected (Isoya 2020). Dense, lush, and teeming with life, it is difficult not to romanticize replicating its success elsewhere.

Pachacamac, however, is not Tokyo, and the absurdity of the initial vision was quickly recognized. The coast of Peru where Pachacamac lies is one of the most arid areas on the planet.

Pachacamac Archeological Sanctuary.

In Lima's historic centre, mean annual precipitation is a mere 24 mm (0.94 in), while other parts of the coastal desert may receive no measurable rainfall for over a decade (Japan International Cooperation Agency 1988). Clearly, a forest planned for a city that receives nearly 1,600 mm (63 in) of rainfall per year would make no sense in the hyper-arid coast of Peru. Despite the false start, there was a conviction that the Sanctuary needed to be surrounded by trees. The social and ecological benefits of urban forests would be an extraordinary way to provide more visibility to an otherwise barren site than a stand of trees. All that was needed was the right kind of trees.

Huarango dry forest

Tree species such as *Acacia macracantha*, *Schinus molle*, *Parkinsonia praecox* and *Tara spinosa* are well adapted to the climate of Peru's coastal desert. However, trees of the genus *Prosopis* – in particular *P. limnensis* and *P. pallida* – stood out from the rest. These trees have been invaluable to the cultures of coastal Peru both as a readily available resource and as keystone species within the local ecosystem. Known as *algarrobo* in northern Peru and *huarango* to the south, *Prosopis* trees epitomize the reputation attributed to desert trees as "islands of fertility" (Schade & Hobbie 2005), due to their capacity to support a diversity of flora and fauna. The *huarango* also prevents soil erosion from wind and water; breaks up hardpan, while increasing soil porosity and water infiltration; moderates temperatures under its canopy with shade; enhances organic matter in soils by dropping leaf litter; and provides water below its canopy through fog drip. The *huarango* can even pioneer degraded soils – saline, alkaline and devoid of humus – and gradually improve them to be able to support plants with higher sensitivities. They are extremely hardy and can live to be over a thousand years old (Beresford-Jones 2011).

  Despite their inherent resilience, *Prosopis* have been unable to withstand the opportunism of humankind. The same trees that survived droughts, floods and multiple cultural upheavals are being cut down in a matter of minutes to be turned into charcoal. In the valleys along the central and southern coasts of Peru where the seventeenth-century Jesuit missionary Bernabé Cobo once wrote "the *algarrobos* or *guarangos* are the most abundant" (Beresford-Jones 2011, 119), they are on the brink of extinction. Deforestation is difficult to reverse especially in arid climates, where cutting down trees triggers a positive feedback loop that exposes soils to desiccation and erosion and eventually creates degraded conditions hostile to most forms of life. History is littered with devastating consequences of deforestation. In *Mountains of the Mediterranean World,* J. R. McNeill famously cataloged numerous examples from around the Mediterranean recounting the same general plot line: elimination of tree cover, loss of soil, irreversible environmental degradation, and eventually economic collapse, social unrest, and political upheaval (McNeill, 1992). In *The Lost Woodlands of Ancient Nasca*, Beresford-Jones argues that in parts of the vast and barren Ica Valley on the southern coast of Peru, it is yet again the loss of woodlands caused by people that led to the demise of its most famous inhabitants (Beresford-Jones 2011). Thus, the objective to grow a *Prosopis* forest around the Sanctuary became not merely an afforestation project but also a restoration project.

Capturing fog to restore a cloud forest

To grow new trees – even trees adapted to arid climates – requires water, however, and a 2019 World Bank report revealed that if climate change were to reduce overall rainfall in the Andes Mountain range where Lima gets most of its water, the status quo would be inadequate to meet demand. Existing water infrastructures of the Lima Metropolitan Area where the Sanctuary is located are just able to meet current needs, and any reduction in supply – more frequent and/or longer periods of drought – could endanger the water security of the region's ever-expanding population (Groves et al. 2019). With population expected to exceed 12 million in 2030 (United Nations n.d.) and with water from conventional sources in such short supply, how can one justify taking water that could be utilized by people to irrigate trees instead? Can a forest be restored without irrigation?

Santa Rosa Island is one of eight islands off the coast of southern California comprising the Channel Islands archipelago. In 1980, Santa Rosa Island together with four other islands became the Channel Islands National Park. After over 150 years of intensive ranching and commercial hunting, Santa Rosa Island became the focus of an ecosystem restoration programme by a team led by USGS and NPS. Since 2015, the team of ecologists, teachers, students and volunteers have been working in a twelve-acre area called Soledad Ridge to actively rebuild the foundational components – soil, vegetation and water – needed to grow a new cloud forest. Soil is being rebuilt by utilizing wattles, wire fences and check dams to prevent erosion, trap sediment and capture leaf litter. A nursery established and tended by the USGS team grows plants to be transplanted at Soledad Ridge from seeds harvested on-site. For water, the team has turned to fog.

In a cloud forest, canopy trees will drip water from fog to create their own precipitation. Fog drip not only provides water for the trees themselves but also nurtures understory plants that serve as nurse plants for tree progeny. In the absence of mature trees, the USGS team designed and constructed fog fences and fog hats from galvanized steel mesh, polyvinyl shade cloth and steel

Fog fences at Santa Rosa Island off the coast of California.

reinforcing bars to serve as fog trapping proxies for nourishing both emergent nurse plants – in this case, *Baccharis pilularis* and *Stipa pulchra* – and newly planted tree seedlings. As the plants mature, they become their own 'fog catchers' able to drip fog on their own and eliminating the need for the fences and hats.

Like Santa Rosa Island, what Pachacamac lacks in rain, it makes up for in fog. From June to December – the austral winter and spring – a coastal fog known locally as *garúa* blankets the Peruvian coast. Although it dissipates to some degree between December and April, fog continues to exert its presence at night and through the mornings, making it a dependable source of water year-round. Capturing fog to cultivate a *huarango* dry forest became the genesis of the design research.

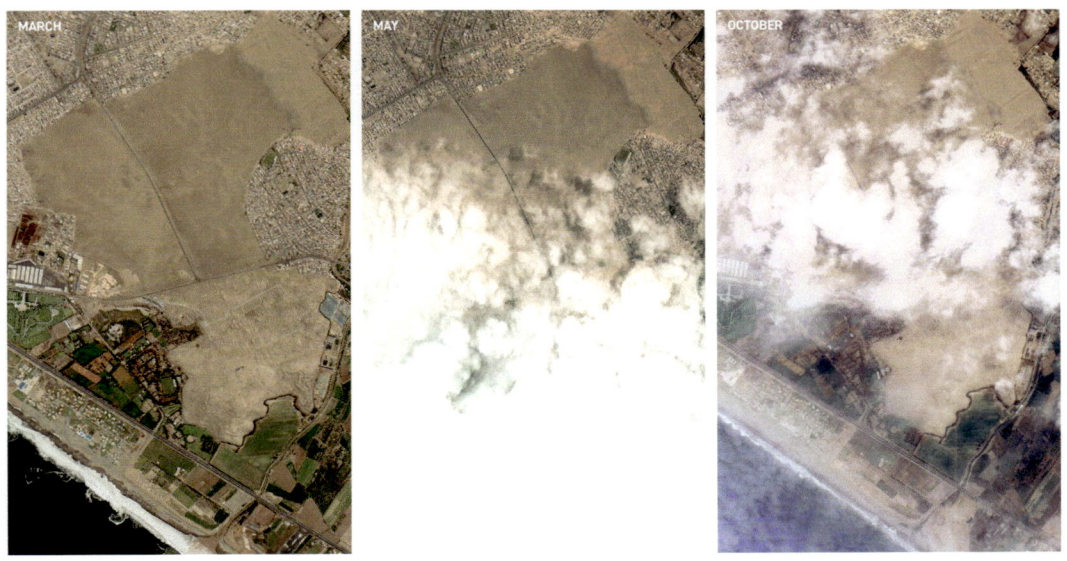

Pachacamac Archeological Sanctuary in various degrees of fog.

A framework for growing their own forest

In the United States, rather than simply saying that they 'plant' trees, landscape professionals often say that they 'install' trees. Implicit in the wording is that the process is finite with completion its primary objective. Although the notion of 'installing' trees may have its merits, a forest cannot and should not be installed because a forest can only be grown. By growing the forest from scratch, one is able to grow a resilient forest since each tree will have lived through the test for the survival of the fittest and would be acclimated to the site. The proposal for the Sanctuary was, thus, not so much a finished design but rather a long-term strategy for growing a forest.

Growing an urban forest provides not only the ecological benefits for which they are perhaps most well-known but has the potential to bring together disparate groups to promote inclusivity and a sense of belonging (Konijnendijk, 2018). In 2006, the Royal Botanical Gardens Kew with support from UK's Darwin Initiative began the "Habitat Restoration and Sustainable Use of Southern Peruvian Dry Forest Project". The project – also known as *Proyecto Huarango* – took an integrated ecosystem approach focused on three main objectives: knowledge recovery, biodiversity recovery and connecting plants and people. The project was successful on multiple fronts, and accomplishments include: forging meaningful relationships instrumental to ensure project longevity; creating educational opportunities for local students; establishing essential infrastructure, such as nurseries for propagating endemic plants from locally harvested seed; expanding knowledge of the local ecosystem; and creating local interest and awareness for the conservation, restoration and sustainable management of *huarango* dry forests (Whaley et al. 2010).

The logistical framework for the design proposal was largely modelled on this project. The possibility of orchestrating *Proyecto Huarango's* involvement of local schools in all aspects of the process – propagation of plant material, planting, watering and maintenance, data collection, disseminating information and educating the public – was fascinating. The propagation of endemic plants was imagined as being integrated into the education programmes at local schools and at the existing Pachacamac Site Museum. By cultivating plants from seeds collected from local stock, the specimens are ensured compatibility with the site. The project proposes to engage schools in the planting process turning it into a rite of passage for every student. Each student would plant several seeds when they first enter elementary school and participate in a planting festival upon graduation. Over time, many of the plants in the park would be associated with a specific individual.

The strategy to evolve this landscape requires time scales that stretch ten, twenty, fifty, to a hundred years into the future. Being both a participant and witness to the emergence of a totally new forest landscape provides a different sense of gratification – one that is more meaningful and long-lasting – from one provided by 'instant' landscapes with which we are all too familiar. Those who have taken part in the planting will reminisce how it came to be, and those unaware of its past will perhaps see the cultivated landscape as a natural condition. The project's ambition is for the forest and the Sanctuary together to eventually become a part of the collective memory of the Peruvian people.

Fog fence concept.

First a fence – gradually, a forest

There is, however, an obvious downside to growing a forest from the ground up, and that is quite simply the time it takes for the new forest to emerge. At Downsview Park in Toronto, for example, the multinational design firm, OMA, proposed a flexible framework that relied on trees rather than architecture to give identity to various programmatic spaces within the park. The growth of trees being on a different timescale from construction, many people have been frustrated by the lack of visible change at the site (Toronto Star, 2014). Unlike a forest-like landscape installed by builders in a matter of days, a forest that has yet to grow looks barren at worst and sparse at best.

The physical implementation of our design proposal begins with the installation of fog catchers throughout the site. The fog fences would nurture seedlings of *Pluchea chingoyo*, *Scutia spicata* and *Vallesia glabra*, which would in turn serve as nurse plants for *huarango* seedlings. However, unlike those built purely for utility, we envisioned the fog fences at the Sanctuary being more than mere surfaces to capture fog. In the spirit of 1960s and '70s land art projects of the western United States, we imagined them to act as visible markers on the land for the dry forest restoration underway and for delineating the territory of the Sanctuary. As demonstrated on Santa Rosa Island, the equipment for collecting water from fog can come in various forms. For the Sanctuary, we

 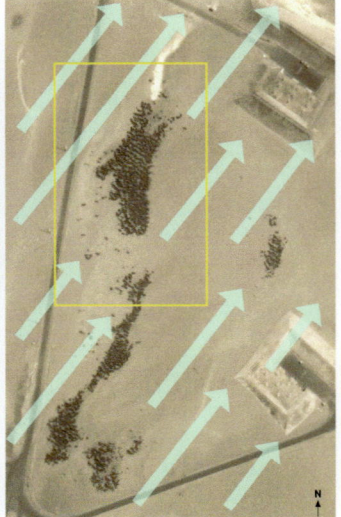 

Bands of *tillandsia* reveal the direction of fog movement.

imagined the fog catchers to be a series of short fences relatively low to the ground (0.5m to 1m) – short since fog fences are not meant to enclose or exclude, and low in order to avoid capturing too much wind. The orientation and alignment of the fog catchers were determined to maximize water collection.

Over time, the fog fences would become obsolete. Plants will mature to capture water on their own, and in what was once ambiguous territory, an emergent forest will define a new zone for activity and interaction. People and forest would be inextricably linked because every tree in the new park will have been planted by generations of children in the surrounding communities. Thus, a framework for co-creation would culminate in a green infrastructure to benefit its closest neighbours.

Bird's eye views of proposed fog fences and emergent *huarango* forest.

Plan diagram of proposed new forest.

Sections along the new linear forest.

Fog capturing pavilion.

View of existing and proposed conditions at the edge of the Sanctuary. The proposed view shows an emergent *huarango* forest adjacent to a new walkway and an existing playground.

Forests as infrastructure

The proposed forest would benefit the Pachacamac Archaeological Sanctuary and its surrounding communities in various ways including but not limited to creating a scenic backdrop to an important cultural site, providing residents with a place of respite, cultivating a living collection to introduce residents and visitors to endemic species, and becoming a source of knowledge for habitat creation. In order to cultivate a forest without further depleting an already degraded water supply, the proposed strategy takes advantage of the region's abundant fog. The fog-capturing fences would not only transfer water from the air into the ground but also for a while serve as territory delineating markers on the land. Its eventual obsolescence would indicate the reemergence of a vibrant urban forest to celebrate one of the cradles of civilization.

In a world that continues to wipe out mature forests faster than the rate at which new forests can be established, mitigation, conservation and restoration are no longer adequate. With the tangible effects of climate change upon us, the self-destructive approach of short-term maximization must be abandoned in favour of long-term visions for new kinds of urbanisms that are net-positive in fundamental resources such as clean air, water, soil and forest cover. Past civilizations did not know of the devastation they were sowing when they denuded landscapes of forests for agriculture and grazing. Today, ignorance is no longer an excuse and growing forests even in the most unlikely conditions is a matter of survival. Over time, forests need to be perceived not as a resource to be used and depleted but something more akin to infrastructure – a fundamental component of our cities that need to be built, expanded and maintained.

Acknowledgement

The design proposal for the Pachacamac Park design competition was the product of a collaborative effort involving students of the University of Southern California's Master of Landscape Architecture programme: Dani Cong, Ziran Ling and Emma Jinyi Wang; Tetsuo Kondo Architects Tetsuo Kondo and Mika Yamada; a+L Mio Watanabe; and Tajima Open Design Office (TODO) Takako Tajima. Unfortunately, we did not win the competition. In 2020, the winning entry – "El Zócalo y El Manto" by Tomás McKay, Kushal Lachhwani and Pablo Alfaro – is in design development.

References

Beresford-Jones, David G. (2011). *The Lost Woodlands of Ancient Nasca.* Oxford: Oxford University Press.

Groves, David G., Bonzanigo, Laura, Syme, James, Engle, Nathan L., and Rodriguez Cabanillas, Ivàn. (2019). *Preparing for Future Droughts in Lima, Peru: Enhancing Lima's Drought Management Plan to Meet Future Challenges.* Washington, DC: International Bank for Reconstruction and Development / The World Bank.

Isoya, Shinji. (2020). Creating Serenity: The Construction of the Meiji Shrine Forest. July 8, 2020. https://www.nippon.com/en/japan-topics/g00866/

Japan International Cooperation Agency. (1988). *Final Report for the Master Plan Study on the Disaster Prevention Project in the Rimac River Basin: Main Report.* https://openjicareport.jica.go.jp/617/617/617_709_10652121.html

Konijnendijk, Cecil C. (2018). *The Forest and the City: The Cultural Landscape of Urban Woodland Second Edition.* Cham: Springer.

Mann, Michael E., & Gleick, Peter H. (2015). Climate change and California drought in the 21st century. *PNAS*, 112, 13, 3858–3859. https://doi.org/10.1073/pnas.150366711

McDonald, Robert I., Weber, Katherine, Padowski, Julie, Flörke, Martina, Schneider, Christof, Green, Pamela A., Gleeson, Thomas, Eckman, Stephanie, Lehner, Bernhard, Balk, Deborah, Boucher, Timothy, Grill, Günther, Montgomery, Mark. (2014). Water on an urban planet: Urbanization and the reach of urban water infrastructure. *Global Environmental Change*, 27 (2014), 96–105. https://doi.org/10.1016/j.gloenvcha.2014.04.022

McNeill, J.R. (1992). *The Mountains of the Mediterranean World.* Cambridge: Cambridge University Press.

Nowak, David J., & Dwyer, John F. (2007). Understanding the Benefits and Costs of Urban Forest Ecosystems. In Kuser, John E. (ed.), *Urban and Community Forestry in the Northeast,* 2nd ed. (pp. 25–46). Dordrecht: Springer. https://doi.org/10.1007/978-1-4020-4289-8_2

Pataki, Diane E., Alberti, Marina, Cadenasso, Mary L., Felson, Alexander J., McDonnell, Mark J., Pincetl, Stephanie, Pouyat, Richard V., Setälä, Heikki, Whitlow, Thomas H. (2021). The Benefits and Limits of Urban Tree Planting for Environmental and Human Health. *Frontiers in Ecology and Evolution*, 9 (April), 1–9, 603757. https://doi.org/10.3389/fevo.2021.603757.

Pincetl, S., Gillespie, T., Pataki, D.E., Saatchi, S., and Saphores, J. D. (2013). Urban Tree Planting Programs, Function or Fashion? Los Angeles and Urban Tree Planting Campaigns. *GeoJournal* 78, 3, 475–493. https://doi.org/10.1007/s10708-012-9447-x.

Schade, John D., & Hobbie, Sarah E. (2005). Spatial and temporal variation in islands of fertility in the Sonoran Desert. *Biogeochemistry*, 73, 541–553. https://doi.org/10.1007/s10533-004-1718-1

Schemenauer, Robert S. & Cereceda, Pilar. (1991). Fog-Water Collection in Arid Coastal Locations. *Ambio*, 20, 7 (Nov., 1991), 303–308.

Schemenauer, Robert S. & Cereceda, Pilar. (1994). Fog collection's role in water planning for developing countries. *Natural Resources Forum*, 18, 2, 91–100. https://doi.org/10.1111/j.1477-8947.1994.tb00879.x

Toronto Star. (2014). Development of Downsview Park in Toronto has been a huge disappointment: Editorial. Published April 24, 2014. https://www.thestar.com/opinion/editorials/2014/04/24/development_of_downsview_park_in_toronto_has_been_a_huge_disappointment_editorial.html

United Nations. (n.d.). The World's Cities in 2018. Accessed May 14, 2022. https://www.un.org/en/development/desa/population/publications/pdf/urbanization/the_worlds_cities_in_2018_data_booklet.pdf

USGS Western Ecological Research Center. (2021). Building a Cloud Forest From the Ground Up. October 6, 2021. https://wim.usgs.gov/geonarrative/cloudforest/

Whaley, Oliver Q., Beresford-Jones, David G., Milliken, William, Orellana, Alfonso, Smyk, Anna, and Leguía, Joaquín. (2010). An ecosystem approach to restoration and sustainable management of dry forest in southern Peru. *Kew Bulletin*, 65, 4, 613–664.

VII. FOREST LOGICS, LENSES AND ORDERS. TOWARDS A CLIMATE-FORWARD FOREST URBANISM ALONG THE ERIE CANAL NATIONAL HERITAGE CORRIDOR

Jamie Vanucchi, Maria Goula

The Erie Canal was built in the nineteenth century to connect the Hudson River near Albany, NY to the Great Lakes in Rochester and Buffalo and to open the bounty of Midwestern resources to urban markets on the East Coast. Although the establishment of the railway a few decades later, followed by highway expansion and the opening of the St. Lawrence Seaway, caused its decline, 80% of upstate New York's population lives within 25 miles of the Erie Canal (NYS Canal Corporation), making evident its lasting urbanizing effect. Today, a mix of some publicly managed lands and a majority of private parcels loosely identifies the Erie Canal National Heritage Corridor in the area of focus, near the Montezuma National Wildlife Refuge to the west of Syracuse. It is a region largely defined by the tension between landscape conditions and the implementation of the Military Tract of Central New York, nearly 8,100 km² of bounty land set aside to compensate New York's soldiers after their participation in the American Revolutionary War. Known as the Ontario Lowlands within the Eastern Great Lakes Lowlands ecoregion, the area was once part of a post-glacial, 80-square-mile wetland complex connecting Cayuga Lake with Lake Ontario (USFWS), later cloaked with forests dominated by *Fagus grandifolia* and *Acer saccharum* (United States Environmental Protection Agency). High agricultural productivity due to deep loamy and muck soils and milder temperatures made the study area an important farming region for the state, after lands were taken from the Cayuga Nation and sold for farms in the late eighteenth century. By 1829 in Cayuga County, 187,495 of 446,150 total acres (40%) were listed as "improved", with 116 saw mills and 52 grist mills in operation (County of Cayuga 1829). Although forests have been returning since at least 1938, climate-affected migrants are expected to boost populations around the Great Lakes Region in the coming decades (Goodell 2018). At the same time, droughts in California and an increasing number of growing degree days and ample freshwater in New York are rumoured to be causing increasing agricultural land speculation in the state. Rising demands for new urban development, solar energy and agriculture mean land use conflicts are likely. Another wave of deforestation, similar to that which occurred in the nineteenth century, is looming. This research acknowledges the important work that forests do to complement and counter human development, including provision of timber and other resources, flood mitigation and erosion protection, animal and plant migration corridors, carbon removal and storage and maintenance of biodiversity. In the late eighteenth and early nineteenth centuries, progress included deforestation, taking territory, movement west, farms anywhere and everywhere, city-building. Into the future, could forests provide a diverse matrix for urban and rural uses which responds to changing landscape conditions and experiments with mutualistic modes of human-nature relations?

Forests as urbanizing infrastructure

Two recent, relevant exhibits about the countryside (Koolhaas & Bantal 2019; Marot 2019), demonstrate a hybridization of the rural, its historic evolutions and possible futures as a diverse productive territory. The four scenarios shared by Marot and curational team in the *Taking the Country's Side* exhibition (2019), "Incorporation: *The capitalist metropolis takes over the countryside,* Negotiation: *Urbanism integrates agriculture as a palette of new components,* Infiltration: *Agricultural and horticultural practices colonize cities and their fringes and* Secession: *New forms of self-reliant communes part with the metropolis*", represent situations, already happening to some extent, that raise the question: how are our futures limited by these speculations? Can the country's elements induce different urban models, beyond grids, blocks, or enclaves where transportation is the main structuring element? In the illustrations of the four scenarios, forests are not considered as structural, although they are undoubtedly part of the countryside. Instead, forests read as edges or forested horizons where grids imposed by geometries of production loosen to give place to informal patterns of low-density inhabitations, where the landform is more complex and the forests, part of the domestication of the wild, give legibility and structure to the productive enclaves between them.

The design research herein shares intellectual affinities with a territory of ideas where the countryside and its people, histories, materialities and futures are foregrounded, as Marot's thorough investigation does. Even so, the investigation at hand focuses on forests as an inherent, ordering element of the countryside. Existing forests in the study area, although mostly immature/successional, are valued as important guides towards a deeper (or maybe just different?) understanding of the landscape and its range of future possibilities. A series of designed forests are informed by these extant woodland patches and hedgerows, and then extend them, seeking to both enhance their capacity to claim ground for the long-term and retain their inherently dynamic nature. Can this combined matrix of existing and designed forests act as a shapeshifting urbanizing infrastructure, to transform abandoned agricultural lands to carbon-sequestering and storing forests, while diversifying local economies, establishing alternative modes of relating to nature, rebuilding the heritage corridor, and expanding the array of recreational experiences accessible to residents and tourists?

Using logics to find forest orders towards future forest types

To understand existing forests along the corridor, structural, change and performance-based logics are understood through their corresponding lenses. Lenses provide unique 'readings' of possible forest orders, each driven by a distinct set of organizing forces. A few examples of this approach are described in more detail in this section, including readings of Erie Canal forests using the geomorphologic, fallow and disturbance lenses.

Logic (conceptual frame)	Lens (used to provide a 'reading' of the forest)	Factors directing future forest orders	Future forest types derived using logic and lens	Species included
Structure-based	Geomorphology of the region (drumlins and muck soil/swamps	In carbon-sensitive times, muck soils demand additional protections, management and possibly collective stewardship	Forest commons, conservation	*Acer rubrum, Acer saccharinum*
	Landscape armatures	Includes both public/private lands, adds trails, heritage focus, degrees of wetness in the form of ponding, flooding and/or hydric soils	Persistent forested structures or forest armatures (also heritage); long-lived and iconic species	*Platanus occidentalis, Metasequoia glyptostroboides*
Change-based	The fallow	Push-pull between farm and forest, fallow types are always in transition, due to projections for climate migration and increased farming and urban development	Transitional forests, agroforestry types (forests as farm), hedgerows, short-term harvest; carbon farming forests	*Populus tremuloides, Populus deltoides, Betula spp., Tilia americana*
	Disturbance (pests, flooding, climate change)	Expanding wetness, declining *Fraxinus* spp., species migration northward, assisted migration	North-south corridors, connected matrix builds resilience	*Taxodium distichum, Liquidambar styraciflua, Nyssa sylvatica*

Logic (conceptual frame)	Lens (used to provide a 'reading' of the forest)	Factors directing future forest orders	Future forest types derived using logic and lens	Species included
Performance-based	Carbon and climate *mitigation*	Assessing forests' varying capacity to sequester and store carbon, carbon markets and additional income for landowners, concepts of additionality and 'adaptive mitigation'	Designed forests for carbon, water quality, floodplain management; aggregated forests of numerous small landowners (collectives)	*Juglans nigra, Pinus strobus, Robinia pseudoacacia, Catalpa speciosa, Liriodendron tulipifera*
	Heritage and climate *adaptation*	Land back, co-management with Indigenous peoples, loss of *Fraxinus* beyond a forestry issue; heritage species and sustainability means considering future generations	New heritage types along canal loop; forests of resistance to guide future development	*Quercus spp.*

Forest logics, lenses and future potential orders. Three conceptual frames, or logics, direct a series of different readings of the forests along the canal. Each lens reveals a distinct forest reading by combining unique sets of data over relevant timescales. Structure-based logics tend towards designed forests for conservation, shared forest commons and forested armatures that provide long-lived, forested structures that add vertical dimension and reinforce the sometimes hidden canal. Change-based logics that centre forest dynamics suggest designed forest orders that work with transitions and disturbance, either by short-term management or by creating connectivity to increase forest resilience and adaptive capacity. Finally, the performance-based logic and associated mitigation and adaptation lenses centre people and forest relations.

The lens of fallow land (we define as land somewhere between idle farmland and mature forest) might seem odd as a mechanism for understanding forests, but in the northeastern U.S., land left idle often quickly returns to forest, so the fallow lens reveals forest change. An example of fallow's push-pull between forest and farm is illustrated by the digging of the canal and subsequent lowering of the water table by up to 10 feet, causing the contraction of swamps and making room for additional agricultural expansion. Tracing hedgerows as an indicator of agricultural intervention, we

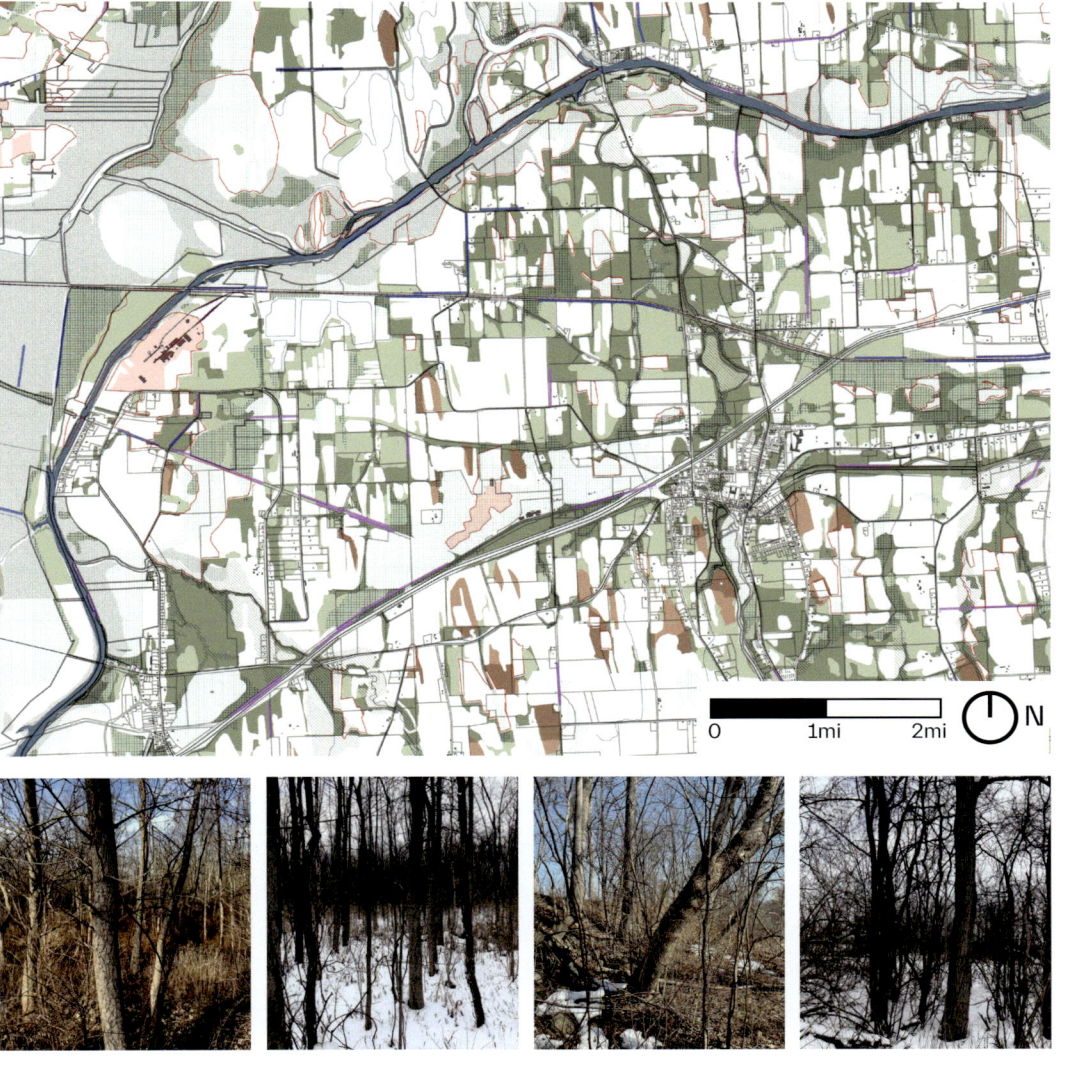

Tracing forest time using the lens of fallow lands. Through the change-based lens of the fallow, we trace the transition from forest to farm and back to forest in the area between the barge canal (large blue line to the north) and the old canal to the south around Weedsport, NY.

find that once a prevalent feature in the study area, hedgerows were either removed as farm conglomerates formed to supply commodity crops to a market extending beyond the region, or sometimes consumed by successional forests in places where farms have been abandoned. A substantial gain in forest cover since 1938 reveals a complicated story of regional farming that likely combines economic and social pressures of small farm life, the loss of the canal as an economic driver and the expanding wetness found in the many low-lying areas along the corridor. Sometimes the trajectory of abandoned farmland's return to forest is slowed by the invasion of aggressive and non-native shrubs such as *Rhamnus cathartica* that form dense colonies and shade out saplings. This shrub-dominated condition can persist for years or even decades, fixing fallow lands in a condition that does not allow forest regeneration. Less obvious are the subtle effects of agricultural histories, features such as subsurface plow pans and disrupted nutrient cycles and mycelium networks that might create a hidden order in current and future forests. These forces may be partly to blame for the lackluster regeneration of canopy species in forests throughout the region (evidence of a re-ordering): one study reported that 38% of study plots in this area had fair to poor regeneration of canopy species and 58% fair to poor for timber species (Shirer and Zimmerman, 2010). In figure 1, the pink is area deforested between 1938 and 2018; red outline is 'old forest', continuous forest since before 1938; light blue is flood-prone area; blue hatch is area prone to ponding; light green areas were reforested between 1938 and 1963; dark green were reforested between 1963 and 2018; dark pink areas are sparsely vegetated, meaning not yet forested (e.g., meadows dominated by *Solidago* or *Rhamnus*, etc.); blue lines are hedgerows added since 1938; purple lines are hedgerows removed after 1938.

  The structural logic of geomorphology is found in soils and fields of drumlins tilted slightly northwest/southeast—the remnants of glacial action in the area during the last ice age around 10,000 years ago. These mounds of mostly loamy and gravely soil have steep side slopes difficult to transverse with typical farm equipment, as well as southwest or northeast aspects, forested with species like *Quercus rubrum*, *Prunus serotina* and *Acer saccharum*. Between the drumlins, clay loams and deep and carbon-rich muck soils support wetland ecotypes, creating much diversity across the landscape section. Muck soils are deep, organic soils with the organic component derived from woody and herbaceous plant materials. These soils are found in pockets throughout the area on nearly flat terrain and have been prized for farming specialty crops (sugar

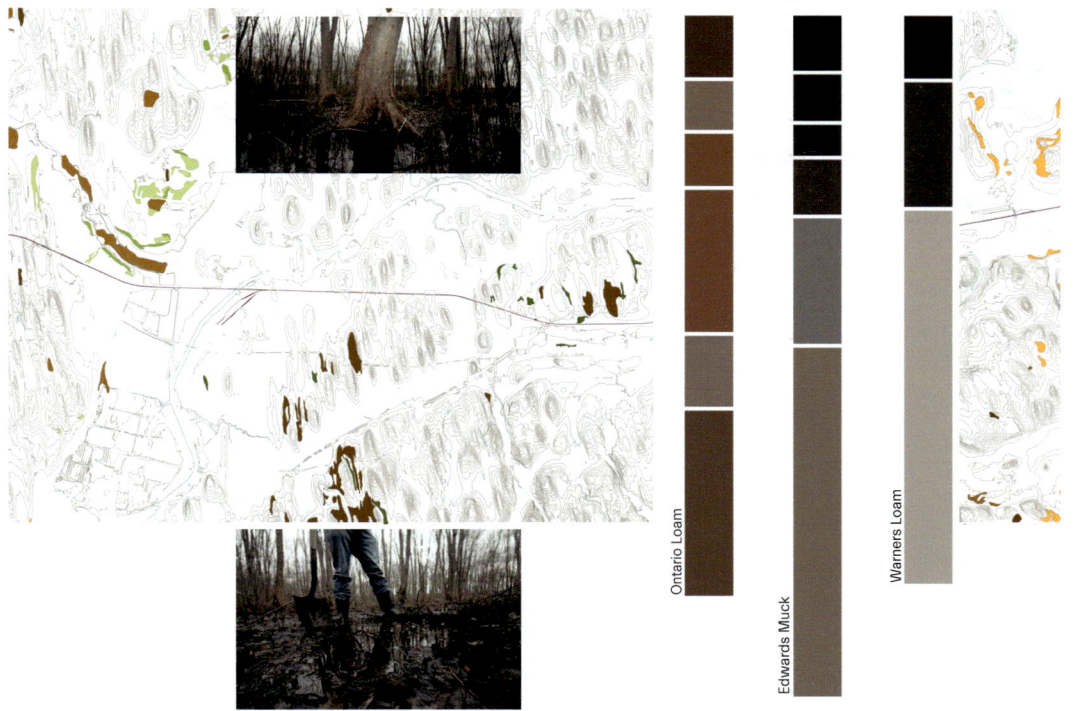

Drumlin geomorphology of the region with mucklands. Carlisle Muck, is depicted in orange and brown. Ontario Loam and Warners Loam are found on drumlin slopes and support species such as *Prunus serotina*, *Quercus rubra* and *Fraxinus pennsylvanica*. Profound differences in soil types across landforms are partly what orders the forest here.

beets, potatoes) for centuries. In places where muck soils have been left undisturbed or returned to wetland, swamps dominated by *Acer rubrum* and *Acer saccharinum* proliferate. Once dried and worked, the muck is vulnerable to subsidence and wind erosion.

Diseases and pests affecting trees are fast-moving, re-ordering forces for forests, seen in the emergence of Dutch elm disease and the loss of *Ulmus americana*, a regionally abundant species later replaced with *Fraxinus* spp. and now in the invasion of the emerald ash borer (EAB). EAB arrived in 2009 and is set to decimate the *Fraxinus* spp. in the region, where 16-28% of the total basal area of trees in the region are black, white and mostly green ash (New York State Department of Environmental Conservation). The loss of ash represents a major disturbance; the tree is a significant actor in the forests in our study area and is important to

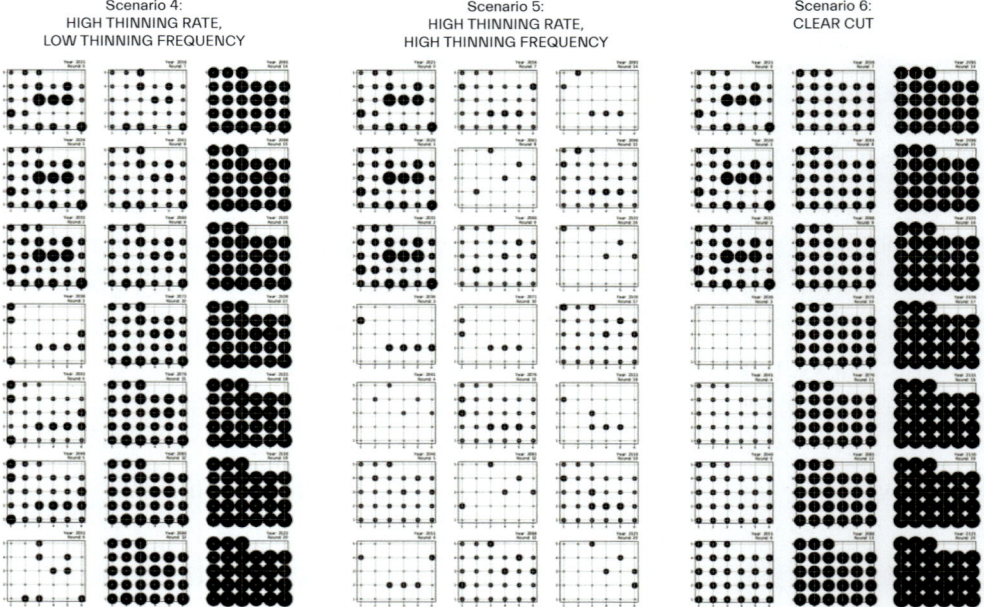

The effects of disturbance on carbon flows in existing forests. Axons depicts number of trees per 50×50 ft field plot. Graphs show total above-ground carbon storage and sequestration rates (in gray) and % contribution by *Fraxinus* spp.. Black and white diagrams explore the affect of silvicultural management strategies on carbon storage.

indigenous basket makers (Wall Kimmerer, 2013), wildlife and is one of New York State's most commercially viable species. In some of our own field study plots, ash trees contribute up to 98% of the carbon sequestered annually, and often ash saplings account for almost all regeneration. This is a huge loss for New York forests, but opens the door for design intervention to both caretake the ash until the EAB threat has passed and to introduce species, such as southern genotypes of existing cohort species, those migrating north but not here yet, or new forest orders. Another impactful disturbance is climate change, more slow-moving than the EAB but accelerated beyond the pace of tree species migrations. In this region, more intense storms and increased precipitation will enlarge areas of wetness (ponding, flooding), expanding resistance to conventional development and agriculture.

Future forest orders

The material realities and trajectories of change for forests along the corridor constructs a multilayered foundation for a forest urbanism that surfaces and responds to some existing orders while introducing others. This dynamically-managed forest matrix engages material structures with multiple logics, providing a framework for placemaking, articulation, connectivity and novel modes of production and inhabitation now and into the future. A future forest matrix anticipates change and builds response to change into forest design and management to foster better synergies between human settlements and forests, needed to mitigate (globally) and adapt (locally and regionally) to climate change. Five forest types that compose the matrix are described below.

Persistent and resistant forest armatures.
Complicating heritage and anticipating change

Structure is found in nested armatures, defined by landscapes of poorly drained or hydric soils, ponding and flooding, persistent because they resist conventional agriculture practice. It is this persistent wetness that characterized the region for thousands of

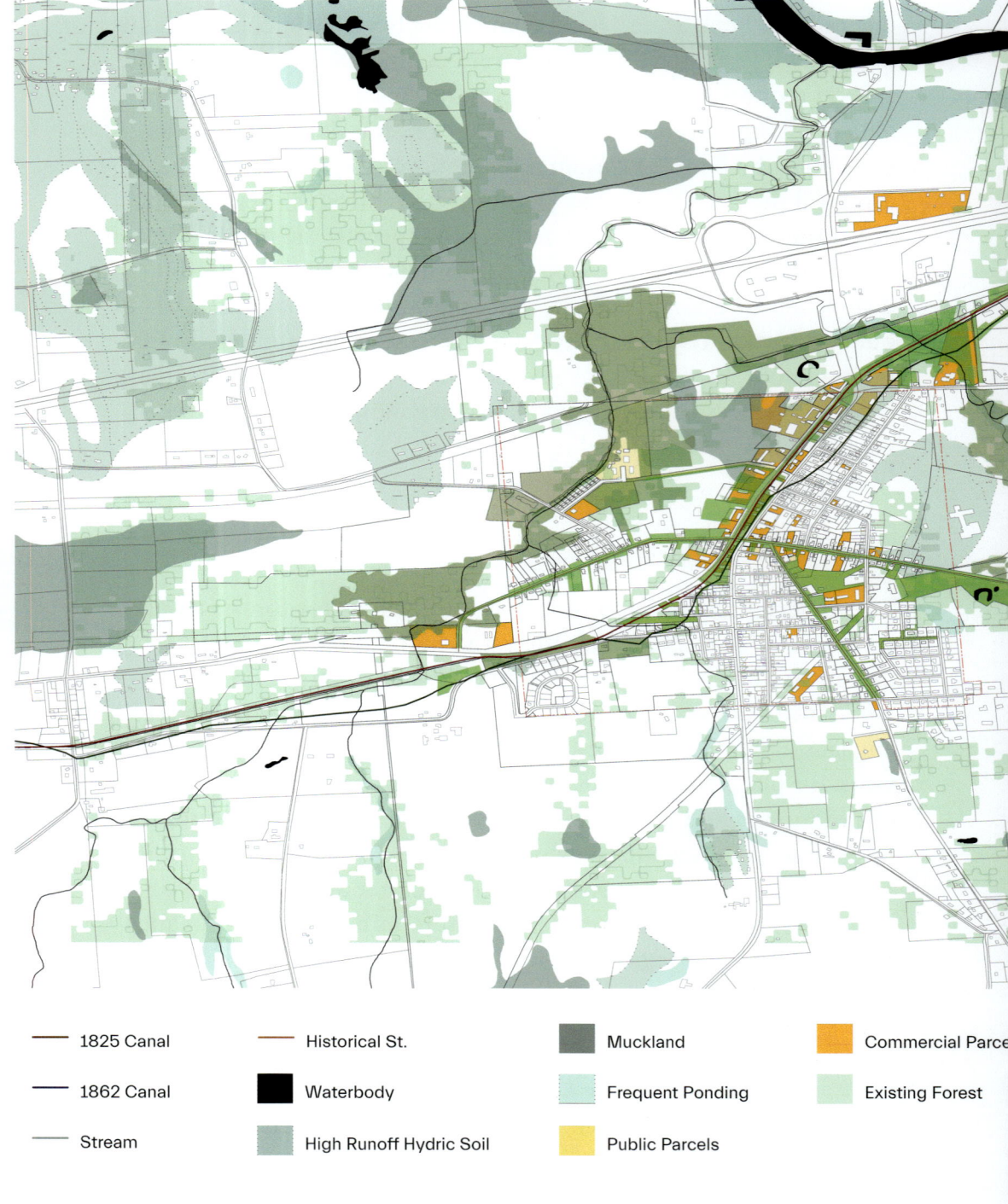

Forest armatures provide an ecological infrastructure for future urban growth.

years and often drives farm to fallow. Other armatures are seen in persistent structures such as the canals and their traces. Landscape-based heritage for the canal corridor is structured around these armatures across scales, as they reflect the memory of performance and cultural use and provide a sustained framework for a forest matrix that guides future housing, recreation and movement and generates better synergies with human settlement. This area shows evidence of 8,000 years of continuous use by Indigenous peoples through the wetlands, floodplains and uplands, and Montezuma National Wildlife Refuge includes Kipp Island, called "one of the most important Native American settlements ever identified in New York" (U.S. Fish and Wildlife Service). Canal heritage must be inclusive of multiple histories and resilient to future change.

Forests armatures are tested in the town of Weedsport, population 1,798 – a typical town founded along the Old Erie Canal. From historical and soil maps we detect a wetness armature along Putnam Brook through a dense riparian forest, lost when crossing between a commercial area and industrial zone before meeting Cold Spring Brook. Our scenario considers providing (where missing) a robust continuity of forest with flood tolerant species to connect the forest corridor with old forest patches on wetlands and muck soils, such as a patch to the east of Putnam Creek, which appears as forest in the 1938 aerial imagery, and also in an 1885 archival perspective drawing (Burleigh 1885).

A second, hidden armature is the first 1825 canal trace, a lost line, either absorbed by fallow forest or eliminated by residential development. This line becomes a heritage armature and allows a different connectivity within the village while complementing the existing, though interrupted, trail of the 1860 modified and expanded Erie Canal. Planting scenarios change in species and density along a hedgerow, depending on the space available and its conditions. Brutus Street and South Street have remained connecting axes, built on the historical paths that gave access to the canal and the rest of territory. One strategy involves planting along these streets with a continuous and initially dense line of trees, including *Platanus occidentalis*, *Populus deltoides*, and *Betula populus*, taking advantage of every available foot. When possible, the line expands into parks, lawns and vacant public parcels to stress the territorial permanence of these pathways and their potential to structure future urban development by gradually creating a heritage alley. Occasionally, the line expands as a forest clump towards areas where farming is still active, yet less viable in the future due to the exacerbating wet conditions.

A different forest logic is tested along the former Erie Canal, now Erie Drive, zoned as a commercial area. Pope and Vasallo (2019) treat vacancy in Detroit as an opportunity to introduce density in an open-ended, spinal spatial pattern in time (a reasonable as well as remarkable outcome of Pope's studies of American cities in the book *Ladders*), showcasing the potential of forests for carbon mitigation, addressed basically as a fill that performs. In the town of Weedsport, NY, we speculate with a spine that is contiguous and incremental, articulating for existing and created forests ecological connectivity, performance through transition to more responsible understandings of site specificity (in this case wetness) and building a rural commons through heritage. Following the concept of forest design as anticipatory armature for residential or mixed-use development, we consider that the pilot in Weedsport can be replicated in many other towns along the canal with similar characteristics. Axons test zoning for forests. We transform around 30% of the parking areas and existing lawns into hedges or forest grids of *Betula populus* and similar species planted 8' on centre. These grids can be harvested strategically to create clearings for the (future) dense mixed-use development, to transform a now commercial zone to a walkable urban area, promoting affordable housing, rentals and tourist accommodations while honoring the historic centrality of this part of town.

Transitional forests in a perpetual state of change
Thinking through the lens of change (and fallow orders), the design of transitional forest types such as post-agricultural forests designed (via species selection and density) to fix depleted nitrogen or break up compacted soils, or short-term harvest forests, allows a shift to temporal strategies for some forest forms. While many forest communities depend on the protection and longevity of landscape

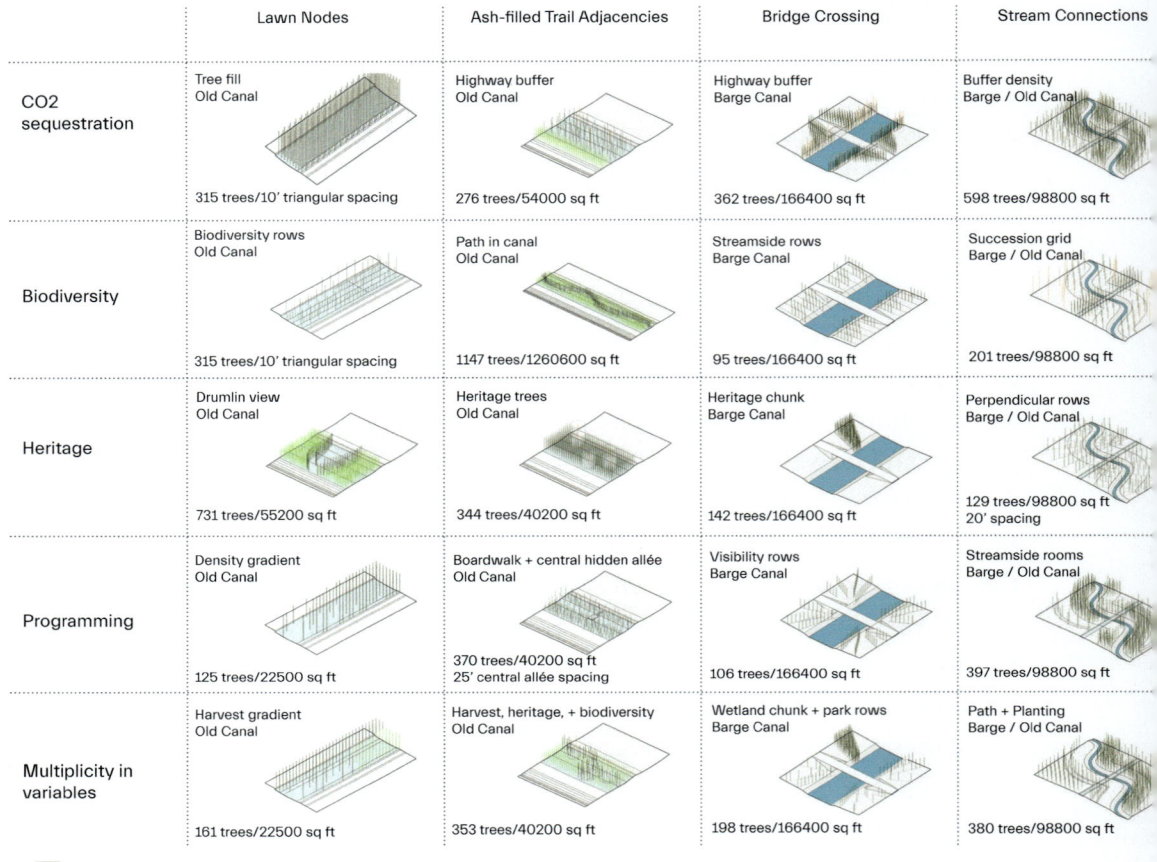

Canal forest typology. Along the Old Erie Canal, a forest typology is developed to address site-specific but repeated conditions while providing an array of performances such as carbon sequestration, biodiversity, heritage and programming. Forested Armatures, transitional forests and forests for connectivity are tested in each condition.

armatures (e.g., red maple swamp), sun-loving and fast-growing species rely on disturbance and occupy the edges between forest and farm. These linear forest spaces sequester carbon in the short term, provide a local source for timber, and could include trails through groves of *Populus tremuloides* and *Betula* spp. designed to encourage resident encounters with the cycles of harvest and succession. Future urban development will have to negotiate responses to the varied forest matrix and, by doing so, generate new urban models.

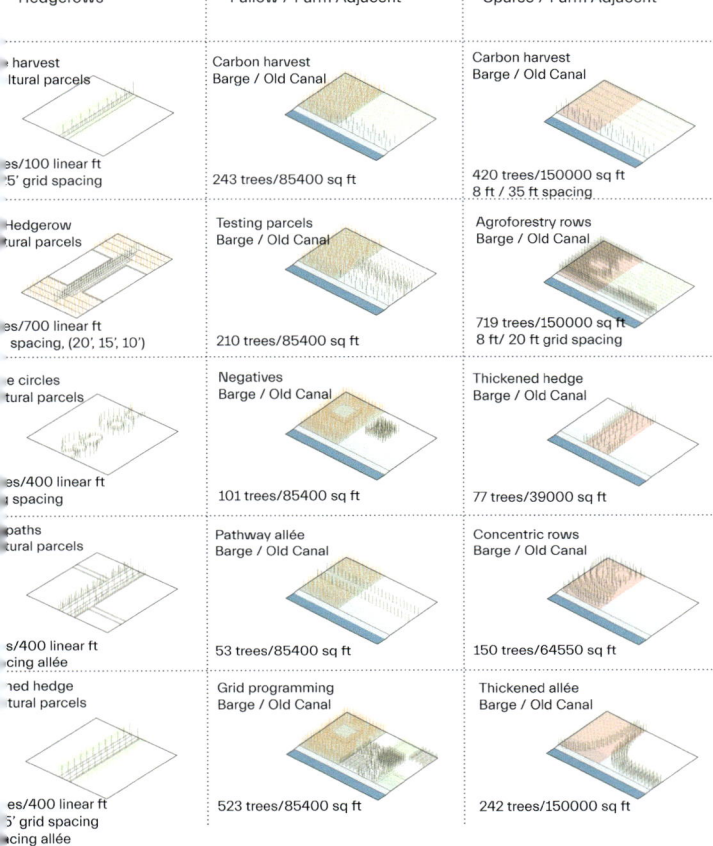

Connectivity forests

Emerging from forest orders driven by *disturbance*, a more resilient and biodiverse forest matrix includes east-west armatures like canals and old railways and north-south creek corridors, to connect across current urban barriers and provide south-to-north corridors for tree migration. The towns and existing development along major roadways can be reforested through acts such as the removal of parking and unconsidered lawns; groves replace single trees; hedgerows connect across patches and property lines. Together, the forests create a material matrix, providing unique spaces and experiences and enhanced livability. Connectivity forests contribute to diversity by articulating space and connecting corridors to old forests across abandoned and extant farmland and sometimes hedgerows, additionally containing a remix of species with one eye to the past (heritage and 'native' species with projected climate resilience) and one to the future(species chosen for their ability to sequester carbon at faster rates, as well as those migrating northwards but not here yet).

Imagining a forest commons and community

In the state of New York, 60% of lands are forested and 74% of forest lands are privately owned (Cornell Cooperative Extension) and managed according to a set of individual property owner values. Managing woodlots requires skills, knowledge, money and time, and can be daunting for woodlot owners. We imagine a forest owners' and stewards' collective that inches us away from individual rights towards the creation of a forest managed across property boundaries and guided by a set of shared values such as beauty, wildlife habitat and harvest for forest health. In a carbon-sensitive future, mucklands demand new conservation mechanisms and collective stewardship that could combine swamps with high-value but sensitive agriculture amidst an agro-forestry matrix. If successful, this collective might experiment with the creation of forest commons, where access is opened for walking and possibly harvest of forest provisions such as fiddleheads, mushrooms and nuts. A commons forest could prioritize the sequestration and storage of carbon for climate mitigation and aggregate smaller properties so the collective can enter the offset marketplace, while moving the local and regional community towards increased adaptive capacity (increased shade, flood mitigation, local timber sources). Expanding our notion of what makes a forest community to *non-human* beings mandates a rethinking of tree 'labour' and harvest. Mycorrhizal associations and insect-tree relations are part of forest health and extend the forest commons to thousands of species that might be considered as part of a community of care. New communities/communions are assembled when alternative plant communities take shape and if private landowners adopt a set of collective practices to connect their lands to adjacent land owners' and public entities along the armature. Regional connections are made along the route, and taking action towards climate mitigation tethers these once isolated towns with larger cities nearby and a global community working to create a more hopeful future.

The research approaches forest urbanism as a frame to establish an inverted hierarchy in the making of urban environments. In the line of thinking of Marot's sub-urbanism (Marot 2003) forests' specific qualities (species, performance, placemaking capacity, landscape heritage) can be an anticipatory armature that could define alternative urban models by adapting them to existing forest orders. This ambitious positioning, in order to become influential for urbanism, asks for increased specificity and awareness of forest logics and orders.

This article focuses on the study of fallow forest logics in the area of Montezuma National Wildlife Refuge along a semi-abandoned trace of the historic Erie Canal. The knowledge and design iterations developed in the research prompt the question: does the study of forest logics lead to novel urban orders? Could existing forest orders and related designed orders suggest a compatibility of density and forest as part of a re-invention of a linear city along the Erie Canal? Could designed forests that can activate and reinforce the existing armatures of the historic Erie Canal could provide a cohesive urbanizing infra-structure for existing fragments of lax urbanity along the corridor? Can the wet, organic soils and the fallow forests that grow on them be understood as a valued land system within the larger matrix, providing a model of responsible productivity, if not a collectively stewarded rural commons?

References

Burleigh, L. R. (Lucien R.); C.H. Vogt & Son; Burleigh, L. R. https://www.loc.gov/item/75694869/. Perspective map not drawn to scale. Bird's-eye-view. LC Panoramic maps (2nd ed.), 654 Library of Congress Web.

Cornell Cooperative Extension. http://ccecolumbiagreene.org/agriculture-and-natural-resources/natural-resources/agroforestry-resource-center/woodland-stewardship, Accessed June 1, 2022.

County of Cayuga: A statistical view according to the latest census. Accessed June 1, 2022. https://www.cayugagenealogy.org/maps/1829/00002893.jpg

Goodell, Jeff. Welcome to the Age of Climate Migration. https://www.rollingstone.com/politics/politics-news/welcome-to-the-age-of-climate-migration-202221/. Accessed August 20, 2018.

Marot, Sébastien. (2019). "Taking the Country's Side" Exhibition.

Marot, Sébastien. (2003). *Sub-urbanism and the art of memory*. London: Architectural Association.

New York State Canal Corporation. Canal History. https://www.canals.ny.gov/history/history.html. Accessed May 21, 2022

New York State Department of Environmental Conservation. Emerald Ash Borer. https://www.dec.ny.gov/animals/7253.html, Accessed May 15, 2022.

New York State Department of Environmental Conservation. New York State Ash (Fraxinus spp.) Distribution Percentage of Ash per Total Basal Area. https://extapps.dec.ny.gov/docs/lands_forests_pdf/ashdistribution.pdf. Accessed May 13, 2022.

OMA (Rem Koolhaas and Samir Bantal). Countryside: The Future. https://www.oma.com/projects/countryside-the-future. Accessed June 2, 2022.

Pope Albert & Vasallo J. (2019). New Corktown, Detroit, MI,USA. In Ibañez, D., Hutton, J., Moe, K. (eds.), *Wood Urbanism, from the Molecular to the Territorial* (pp.338–349). New York: Actar Publishers.

Shirer, Rebecca, & Zimmerman, Chris. (2010). Forest Regeneration in New York State. https://forestadaptation.org/sites/default/files/NYS_Regen_091410_0.pdf. Accessed May 20, 2022.

United States Environmental Protection Agency. Ecoregions of New York. https://gaftp.epa.gov/EPADataCommons/ORD/Ecoregions/ny/NY_front.pdf . Accessed May 20, 2022.

U.S. Fish and Wildlife Service. Montezuma National Wildlife Refuge: Comprehensive Conservation Plan (pp.3–37). https://ecos.fws.gov/ServCat/DownloadFile/44433. Accessed May 28, 2022.

Wall Kimmerer, Robin. *Braiding Sweetgrass: Indigenous Wisdom, Scientific Knowledge and the Teaching of Plants*. (2013). Milkweed Editions: Minneapolis, MN.

ILLUSTRATION CREDITS

All images and drawings in this book are credited to the authors mentioned at the beginning of each chapter, except for the images on the following pages:

17 Duvigneaud, Paul. (1974). *Aménagement du territoire: On a représenté, en regard, deux façons d'utiliser le même paysage.* Ink on paper. *La Synthèse écologique*, p. 2-3.

21 Hallé, Francis. (undated). *Forest Profile*. Ink on tracing paper. 162.5x145 cm.

23 de Briançon, Jean. (1422). *Figure de la vallée de Château-Dauphin*. Parchment. 70x100 cm. Archives départementales de l'Isère, B 4496.

26 Rosenblatt, Vincent. (2012). *Watoriki, Yanomami village*. Photograph. 3543x2357 pixels.

28 Rodríguez, Abel. (2013). *The Chagra Cycle (1 Year)*. Ink, graphite, and watercolour on paper. 50 x 70cm Tropenbos Colombia.

31 Dodds, Kieran. (2018). *Hierotopia*. Photography. 5906x4430 pixels.

33 Juliuz, Clemente. (2018). *Untitled*. Ink on paper. 3460x2440 pixels. Artes Vivas Collection, Verena Regehr.

35 Uccello, Paolo. (1470s). *The Hunt in the Forest*. Tempera, oil and gold on panel. 73.3x177 cm. Wikimedia Commons. https://commons.wikimedia.org/wiki/File:Hunt_in_the_forest_by_paolo_uccello.jpg, accessed May 2024.

38 Prate, L. [J]. (1901 – 1905. *Foresters at Work Mississippi*. Photograph. 4112x3338 pixels Records of the Forest Service, 1870-2022. *https://catalog.archives.gov/id/7001189?objectPanel=transcription*, accessed May 2024.

41 Atelier du Maître de Boèce. (1410-1415). *Les pays de la terre*. Parchment. 4996x6981 pixels. Bibliothèque nationale de France. https://images.bnf.fr/#/detail/931549, accessed May 2024.

43 Ministry of Information and the Arts (MITA). (1963). *Prime Minister Lee Kuan Yew planting a mempat tree at Farrer Circus*. Photograph. 2048x3072 pixels. Ministry of Information and the Arts Collection, courtesy of National Archives of Singapore.

46 City of Oslo. (2012). *General city plan for Oslo 1950*. Ink on paper 1239x1752 pixels. Wikipedia. https://en.wikipedia.org/wiki/File:Generalplan_1950_parker_og_turveier_s_63.pdf, accessed May 2024.

49 Adams, Thomas. (1928). *General Plan of the Park System for New York and its Environs*. Ink on paper. 21.25 x 19.5 inches. Regional Plan Association.

51 Bronder, H. (1954). *Die geschlossene Großgrünpflanzung in der Wohnsammelstraße geht zur lockeren Pflanzung in der Wohnstraße über. Die Bäume gruppieren sich um das Haus. Der Siedlergarten bleibt baumfrei.* Ink on paper. 1841x1194 pixels. *Großgrüngestaltung und Städtebau*, p. 48.

53 Nix, Charles Thomas. (1949). *Bijdrage tot de Vormleer van de Stedebouw in het Bijzonder voor Indonesië*. Ink on paper. 1554x1957 pixels. [PhD dissertation TU Delft]. Heemstede: De Toorts, p. 120.

55 Weller, Richard, Hoch, Claire, and Huang, Chieh. *Deforestation*. (2014). Digital Media. 8875x4974 piexles. *Atlas of the End of the World*.

57 Holten, Katie. (2019-2020). *Forest*. Ink on paper. 2880x3696 pixels. *Katie Holten*.

59 Photograph. *California National Guard, October 11, 2017*. CC BY 4.0 Deed | Attribution 4.0 International, https://www.flickr.com/photos/thenationalguard/37007633253/in/album-72157661451525598/, accessed May 2024.

61 (Left) Williams, Gerald W. (2013). *The major planting areas of the Shelterbelt Project from 1933-1942*. Ink on paper. 1837x3176 pixels. U.S. Forest Service. https://en.m.wikipedia.org/wiki/File:Shelterbelt_Project_planting_areas.JPG, accessed May 2024; (Right) Joseph, Dusek. (1936). *Plains farms need trees. Trees prevent wind erosion, save moisture … protect crops, contribute to human comfort and happiness*. Poster. Silkscreen on posterboard. Chi[cago]: Illinois WPA Art Project [between 1936 and 1940]. https://www.loc.gov/item/98517930/, accessed May 2024.

64 (Left) Ishigami, Junya. (2019). *Tree canopy projection*. Ink on paper. 3508x3508 pixels. In *Another scale of architecture* (p. 58). Tokyo: LIXIL Publishing; (Right) Ishigami, Junya. (2019). *Survey diagram, July 15, 2010 S=1:400*. Ink on paper. 3508x3508 pixels. In *Another scale of architecture* (p. 59). Tokyo: LIXIL Publishing.

66 Desvigne, Michel. *Nîmes Mas Lombard Development*. (2020). Physical model. Michel Desvigne Paysagiste.

78 (Top and Bottom) Johan Östberg; (Middle) Cecil Konijnendijk.

79 Image reproduced with permission from UNECE (2022).

83 (Top) European Forest Institute; (Bottom) Bart Muys.

85 Sven Lorenz.

90, 91 URI staff.

103 (Top and Bottom) Michiel De Cleene.

128, 129, 130 (Top and Bottom), 131 Diane Snape.

143, 154 Kollektif landscape, Fallow.

146 Author processing of the PRDD.

148 Author processing based on BruGIS.

160 Author from Google Earth, accessed 19.8.2023 with images from fieldwork July 2023.

176 Winnipeg Comprehensive Urban Tree Strategy Draft. City of Winnipeg. 2022. p.30.
178 (Top) Atuhairwe, Calvin. Sylvan City Studio: *Urban Oasis*. University of Manitoba, 2019; (Bottom) Dick, Lyle. The Greening of the West: Horticulture on the Canadian Prairies, 1870-1930, Winnipeg, 1945. http://www.mhs.mb.ca/docs/mb_history/31/prairiehorticulture.shtml. Image source: Archives of Manitoba.
180 (Left) Warkentin and Ruggles. *Historical Atlas of Manitoba*. map 74, p. 188) Image: Courtesy of University of Manitoba (Archives & Special Collections).
180 (Right), 181. Evan McPherson. *Fluid Relations: Reframing water on the edge of the Red River*. Winnipeg: University of Manitoba [Master of Landscape Architecture Practicum], 2020.
186 Claudia Lucia Rojas-Bernal, 2024.
190, 194, 197 Barranquilla Verde Archives, 2021-2022.
191 Serrano-Aragundi & Rojas-Bernal, 2024.
206 Dieter Schwerdtle, Hans-Ulrich Plaßmann, 1982. In Informationsflyer 2014, Stiftung 7000 Eichen, Zur Förderung der weltweit einmaligen Sozialen Raum-Zeit-Skulptur von Joseph Beuys. http://7000-eichen.de/fileadmin/Resources/Public/Files/Informationsflyer.pdf.
210 Carsten Pietzsch, 2013. This file is licensed under the Creative Commons Attribution 3.0 Unported license. https://commons.wikimedia.org/wiki/File:Kaputte_Dinosaurier_Spreepark.JPG.
211 Grün Berlin, 2022. https://www.spreepark.berlin/konzept-und-grundlagen/vision-ziele/.
215 Mark Green/Shutterstock.com
218 Kathryn McEachern, 2016.
219 Google Earth, 2022.
221, 224 (Top and Bottom) TODO, 2019.
222 (Left) Michael Douglas, Ph.D.
223 TODO, 2022.
224 (Middle) Tetsuo Kondo Architects, 2019.
235 Authors with Daniel Meyer and Gengjiaqi Chang.
236 Authors with Jianing Zhou; management strategies diagram by Cheng Fei.
238-239 Authors with Xiaochang Qiu.
242-243 Authors with Dominic Malacaman.

COLOPHON

The publication of this book was made possible thanks to the financial support of the International Center of Urbanism, Department of Architecture, Faculty of Engineering Science, KU Leuven, and the KU Leuven Fund for Fair Open Access.

Acknowledgements
The editors would like to thank the co-chairs and participants of the Urban Forests, Forest Urbanisms & Global Warming International Conference, held in Leuven 27-29 June 2022 as well as additional practitioners who agreed to contribute to the Forest Urbanisms Project section. The conference itself benefitted from support by the Department of the Environment of the Flemish Region, Foundation Francqui and UCLouvain as well as by Laura Calders and numerous members of the OSA research group and MaHS/ MaULP programmes. Khalda Eljack extended support with communication and gathering illustration copyrights. The editors also wish to thank the blind peer reviewers for the constructive comments and the team at Leuven University Press.

Editors
Bruno De Meulder
Kelly Shannon

Contributing authors
Marlène Boura
Björn Bracke
Chiara Cavalieri
Koenraad Danneels
Swagata Das
Bruno De Meulder
Rik De Vreese
Kamni Gill
Maria Goula
Cecil Konijnendijk
Colleen Murphy-Dunning
Bart Muys
Alejandra Parra-Ortiz
Jörg Rekittke
Gina Serrano-Aragundi
Kelly Shannon
Takako Tajima
Jamie Vanucchi

Copyediting
John R. Eyck

Graphic Design
Studio Otamendi, Brussels

Typesetting
Crius Group, Hulshout

Fonts
Everett by Weltkern
Life by Linotype

Published in 2024 by Leuven University Press / Presses Universitaires de Louvain / Universitaire Pers Leuven. Minderbroedersstraat 4, B-3000 Leuven (Belgium).

Selection and editorial matter © 2024, Bruno De Meulder & Kelly Shannon
Individual chapters © 2024, the respective authors

This book is published under a Creative Commons Attribution Non-Commercial Non-Derivative 4.0 License. For more information, please visit https://creativecommons.org/share-your-work/cclicenses/

Attribution should include the following information: Bruno De Meulder & Kelly Shannon (eds), *Forest Urbanisms: New Non-human and Human Ecologies for the 21st Century*. Leuven: Leuven University Press, 2024. (CC BY-NC-ND 4.0)

Unless otherwise indicated all images are reproduced with the permission of the rights holders acknowledged in the illustration credits. All images are expressly excluded from the CC BY-NC-ND 4.0 license covering the rest of this publication. Permission for reuse should be sought from the rights holders.

ISBN 978 94 6270 421 3
eISBN 978 94 6166 571 3 (ePDF)
eISBN 978 94 6166 617 8 (ePUB)
https://doi.org/10.11116/9789461665713
D/2024/1869/36
NUR: 648